THE DIGITAL CHAMPION

THE DIGITAL CHAMPION

Connecting the dots between people, work and technology

Simon Waller

simonwaller

First published in 2016 by Simon Waller in Melbourne, Australia.

Edited by Alexandra Davies

contents

list of illustrations

about the author

Born on the beach and raised on fishing boats, Simon's affinity for technology is somewhat unexpected. He hasn't spent his life studying or working in big tech. Instead, what drives Simon is his deep curiosity for making things work. Better. Faster. Cheaper. His ability to make the abstract accessible is his advantage, he can turn a technophobe into a digital native in just a few clicks.

Self-proclaimed as inherently lazy his intent is to use technology to make his life and the lives of those around him easier so they can get on with the work and play that really matters.

His clients include multinational organisations such as John Holland, AMP, Bank of Queensland and OpenText, industry associations such as CPA Australia and The Law Society as well as government agencies and small and medium sized organisations across Australia. The breadth of impact that digital technology is currently having on organisations means his ideas have become applicable across a huge range of audiences.

But above all else, Simon's most significant endeavor is the Digital Champions Club. A community built around helping people in small and medium sized organisations tap into the opportunities of digital technology. Growing up and working in a family business and as a resource speaker for The Executive Connection Simon has always shown a strong affinity for working with these types of organisations, and in his own words, the Digital Champions Club is the best program he has ever developed.

Simon is also the author of ***Analogosaurus: Avoiding extinction in a world of digital business*** and as part of his personal commitment to showing others how to 'live the dream', he resides on the Mornington Peninsula nestled between the beach and the vineyards...whilst through the wonders of modern technology, only milliseconds from the rest of the world.

acknowledgements

In this book I talk a bit about the power of community in supporting the development of digital champions and I am blessed to be a part of two communities that converge at the point where this book has been written.

The first of those communities is Thought Leaders. This is the tribe that inspires me, supports me and pushes me to do my best work. These are my friends, my mentors and my biggest supporters. They are the busiest people I know yet they always have time for me.

The second of those communities is The Executive Connection (TEC). I have been a speaker for TEC for nearly three years and through them I also receive invaluable support and advice from some of the smartest minds in small business.

As much as I am incredibly appreciative of both these communities in their entirety there are a few specific people that need to be singled out.

Firstly to my (official) mentors Laurel McLay, Craig Cherry and now Jason Fox...it has been a privilege learning from each one of you. You have all helped me learn something incredibly important when I most needed it, I hope I get to do the same thing (officially) for others in the Thought Leaders community in the near future.

Secondly to Mykel Dixon who has helped me reconnect with my best self so I can share it with others...you're freakin' awesome.

Thirdly to my TEC Chairs Harvey Martin, John Beech, Adrain Geering and Guy Mycroft. I am incredibly appreciative of the time you have given me over the last couple of years and it has been a privilege learning from each one of you. Your experience and generosity is a reflection of what makes TEC such a valuable organisation for its members.

And lastly, and in no way least, to my family. To my wife Nomes, thanks for continuing to support me on this journey...even when the hours are a little irregular and you don't always get what I do. And to my girls Miah and Poppy, you bring meaning and purpose to everything I do.

foreword (or why I bothered writing another book so soon)

The fact that I have written this book only one year after writing my first one is as big a surprise to me as it is to anyone else. Writing *Analogosaurus* was a bit of an ordeal. The process took close to a year and much of that time was frustrating, if not downright depressing. It took me nearly two months to work out what I was actually going to write about, another three months to 'forget' how to write like an academic (before *Analogosaurus* the last time I had put 20,000 words 'down on paper' or 'up on screen' was when I wrote my master's thesis back in 2010) and nearly seven months of actual writing.

Now, I never say never, but after it was finally published 12 months ago I had no intention of picking up my keyboard again for at least another year...or three.

But then late last year something happened, and within a couple of months I realised I needed to start work on another book.

Within a couple of days of each other I had run workshops with two of my favourite clients to help them identify potential digital projects. At the end of both of these workshops we had found some significant opportunities to improve productivity and information flow in their organisations. In both cases I was then asked to help implement these opportunities. When I did the maths I knew that it would be cost prohibitive for me to do the implementation, and I also knew I couldn't give both of these clients enough of my time to ensure their success.

But I told the clients that I would think about it and promptly set off for a family holiday in New Zealand. It was there, probably about half way up Mt Maunganui, that I realised that although both of my clients needed help, they didn't really need me to do the implementation. I was the wrong person for the job.

The problem with me, and most other consultants out there, is that we don't truly understand our clients' businesses. It's not that we don't care (most of the consultants I know care a great deal) but if we aren't there every day working in the business our understanding of how it operates will always be limited. We may be able to bring knowledge and expertise to bear on a project, but whether we like it or not we will always be an outsider. We won't understand the current systems and how they are used, we don't have any existing relationships, we won't know who's on board and who will need extra care and attention, and we won't know what has been tried before and what has failed.

What I realised half way up the mountain was that my clients didn't need me running their projects, they needed someone on the inside doing it, but that someone would probably benefit from access to some of my knowledge and expertise. Over the next few days I started formulating an idea that would eventually become the Digital Champions Club, a mentoring and support network for people embedded inside small and medium sized organisations delivering the next generation of digital projects.

When I took this concept back to my two clients they both immediately embraced the idea. Since then, the Digital Champions Club has continued to grow. As more organisations came on board I knew that I would need to start capturing my methodology and ideas in a way that could be referenced and shared by members. Initially this was done via private blog posts and videos shared on The Huddle (the Digital Champions Club's social network) but I eventually realised that something more 'solid' was going to be required.

And that is where this book comes in.

The primary reason for writing this book is to support the members of the Digital Champions Club. I have no doubt that this is the most effective and valuable program I've ever created and I'm completely committed to supporting the digital champions currently enrolled in the program in any way that I can.

At the same time I have also come to realise that there are many other digital champions out there, not enrolled in the Digital Champions Club (at least for now) that also need help and support. I have met many of these digital champions in and around Melbourne but know they also exist in other capital cities and other regional centres around Australia.

Sometimes I find them in clients' offices, but often we connect at edgy Meetup groups or other networking events where the digerati gather. They are there, like I am, to connect with likeminded people, talk about ideas, find out what other people are doing and take some of that knowledge back to their organisations.

Most of these people don't carry the job title 'digital champion'. I dare say the majority don't even have a formal role to act in that capacity. They are just passionate in their beliefs about the opportunities that digital technology offers and the need for their organisations to get on board. Ultimately, this book is for all of our digital champions, whether by title or by action, or by want of their enrolment in the Digital Champions Club.

If I had to identify just one thing that made this book such an easy (and enjoyable) book to write after the pain of the first one, it was in knowing exactly who it would help and exactly how it would help them (OK, technically that's two things, but let's not get too pedantic this early on in the piece).

For all the digital champions out there, I hope this book helps you make a difference in your organisations and gets you the support and respect you deserve.

just a couple of things before we get started

This is a book aimed at doing stuff. At a recent conference I saw a keynote from someone working at Atlasssian and they described the organisation as a do-ocracy, meaning that authority and respect flows to the people who "do shit" (their words not mine). This book is written for doers and contains lots of examples of things you could do or things that have been done.

Please don't think that just because one person has done them that you should do them, too. Everyone's circumstances are just a little bit different and what has worked in one organisation might not necessarily work in another. Also, for every app or platform that is used as an example in this book there are probably another dozen or so alternatives that do a similar job. If you want to find out what they are, do a little Google searching or check out services that provide crowdsourced software recommendations such as alternatives.net.

That being said, I hope that some of the examples in this book are a catalyst to you finding new projects you could implement at your organisations...after all, that is why they have been included.

Now, let's go do some shit!

INTRODUCTION

what is digital disruption?

Sometimes I think 'digital disruption' has become the most overused phrase in business. For many it is hard to distinguish between 'disruption' and good old fashioned everyday 'change'. After all, 'digital' is not new. For most businesses it has been around since at least the mid 1990s, and some might wish to track their use of digital technology back another 30 years before that. So if digital has been around for that long, what's so different about what is happening now?

To understand how the technology driven change we are seeing now (especially the change brought by digital technology) is fundamentally different than the 'normal' change we have seen in the past, there are two factors that we might want to look at. The first is the speed of change, and the second is the magnitude of its impact. Although it is the magnitude of impact that has been most surprising, the speed of change is perhaps the hardest to get our heads around. This is because changes in digital technology are happening in an exponential fashion rather than a linear one.

Humans are pretty good at understanding linear change. With linear change, what we have seen in the past is a good indication of what we will see in the future. It is simple, predictable and takes little cognitive energy to work out. Humans are not so good at getting to grips with exponential change, where what we have experienced in the past is a poor predictor of what we will see in the future. The most common example of exponential change, which many experience on a daily basis, is compound interest. Compound interest is why long term investments earn more than we imagine, or why long term loans (such as a home loan) seem to take longer than we expect to pay off. In both cases it is because of the interest that accumulates on top of the interest from previous years resulting in either faster growth or slower payback depending on which side of the money equation we are. Financial planners and accountants will generally try and wow you with 'the magic' of compound interest, primarily because our little human brains struggle so to grasp the concept of accelerating growth.

One of the best illustrations of how linear and exponential change differ comes from the founder of the Singularity University, Ray Kurzweil. In one example he compared taking 30 steps in a linear fashion and 30 steps in an exponential one. If we take 30 steps in a linear fashion we will get approximately 30 metres, but if we take 30 steps in an exponential fashion (where each step is twice as big as the last) we will get significantly further: about 1 billion metres.

Which is the equivalent of walking to the moon and back again.

Exponential change is the type of change that we are currently seeing with technology. If we take Ray's example and apply it to computing power over the last couple of decades we end up with some rather extraordinary results.

Let's just say our journey started back in the mid 1990s with the launch of Windows 95 (remember that?). If each step represents one year then most organisations are about 20 steps into this digital journey. Over that time we have experienced a LOT of new digital technology including the mass adoption of email, BlackBerrys, iPhones, iPads, Wi-Fi, Skype, Dropbox, Google Docs, Facebook (yes, there was a time before Facebook), LinkedIn, 3D printing, machine learning and the foundations of quantum computing. By anyone's estimation this has been a fairly significant couple of decades but because of 'the magic' of exponential growth this will all turn out to be rather insignificant. If we look at all the change we have seen over the last 20 years, it represents *less than 1%* of what technology will be capable of in the next decade.

LESS than ONE PERCENT[1]!

The digital we experience 10 years from now will be 100 times more powerful than anything we have experienced to date. As I mentioned earlier, this compounding of computing power is the bit that is hardest to get our heads around, but it's the magnitude of impact that has surprised us the most.

In the early days of this digital revolution we had a fairly narrow view of what technology could do, and subsequently, which jobs it would impact. We understood that computers

---

[1] Sorry to keep making a paragraph out of a sentence. Probably not the properest approach to style, but I feel this is fairly significant.

were good at managing numbers and during the 1980s we were happy to see our accounting systems and other calculation intensive processes get digitised. We were comfortable with the deployment of robots in highly repetitive and potentially dangerous manufacturing processes. What we didn't necessarily realise was that there were a whole lot more jobs at risk.

The concerns we have about the digitisation and automation of work have led to some important recent research. The first was completed by researchers at Oxford University a couple of years ago and it showed that nearly two thirds of existing jobs were at risk of automation over the next 20 years.

The second, done by McKinsey, took a more nuanced (and in my view, more valid) approach and found out that the problem is both better and worse. McKinsey shows that digital technology doesn't replace jobs, it replaces tasks. Jobs are way too intricate and complex for a robot to do, but if we break jobs down into tasks then technology can do a whole bunch of them[2]...and we won't be waiting 20 years for it to happen. What McKinsey's research found was that about 60% of current occupations were already at risk, with 30% or more of their activities able to be automated by adapting currently demonstrated technologies.

You would be hard pressed to find a job that isn't in some way being touched by technology. We didn't truly appreciate that whether it be designing products, ordering materials,

---

[2] I would say I told you so but that would be a little immature. Instead I would just point out that you can find a more detailed analysis of this in my first book *Analogosaurus: Avoiding Extinction in a World of Digital Business*...which came out well before McKinsey's report.

taking notes in a meeting, writing emails, giving directions or just about any other part of our jobs, almost all work involves some form of information creation, use or sharing...and that is the type of work that digital technology does best, and often better than us.

History shows us that the exact impact of technology is often difficult to predict. But given that technology will become 100x more powerful over the next decades it will surely have an extraordinary impact on the business landscape and be key to the success or failure of many organisations.

In the words of technology commentator Kevin Kelly: "We're just at the beginning of the beginning of all these kind of changes. There's a sense that all the big things have happened, but relatively speaking, nothing big has happened yet."

the gap between extinction and survival

It is fair to say that based on the speed of change and the magnitude of impact, the term 'digital disruption' is an appropriate one. But what we often lose sight of (at both an individual and organisational level) is that it is not technology that is the threat and it is not technology that we are ultimately competing against. Digital technology just provides a new generation of tools that allow us to compete better than before. The threat we are facing is other organisations, organisations that are using technology better than we are.

Research by Capgemini shows that almost irrespective of industry we are seeing an emerging gap between the technology 'haves' and the technology 'have nots'...and the impact it is having on performance is huge. Capgemini's research shows that organisations that are leaders in technology investment and managing digital transformation significantly outperform their peers. Their research shows that on average digital leaders are 26% more profitable while digital beginners are 24% less profitable than their industry average[3]. This means almost 50% of profit is riding on an organisation's approach to technology.

This significant gap in profitability means that this is not solely about financial performance, it is about an organisation's survival. After all, for how long can an organisation afford to continually make 50% less profit than its competitors and still be in business?

What is also significant about Capgemini's research is that only a small part of this performance differential is driven by revenue (digital leaders generate about 9% more revenue than average). This indicates that there is a big opportunity in using digital technology to increase productivity, improve quality and reduce costs.

At the end of the day, digital is as much an opportunity as a threat. For every disruption we experience in our business, there is an organisation out there disrupting us by using digital technology to improve their processes and their profitability. And those who are doing it well are increasing both profit and market share. The real question is with this type of return at stake why aren't *all* organisations doing more when it comes to digital technology?

I think that the biggest cause of inaction for most organisations isn't the failure to identify the potential threat (although unless they have read the previous chapter, they

---

[3] The Digital Advantage: How digital leaders outperform their peers in every industry (https://www.capgemini.com/resources/the-digital-advantage-how-digital-leaders-outperform-their-peers-in-every-industry)

may not appreciate its speed and magnitude). The big hurdle is that they don't know how to address it. Particularly in smaller organisations there is often the perception that they lack the resources and the knowledge required to take action. This may or may not be true at the moment, but there is definitely no reason for it to be true in the future.

What I want to show you with this book is that there are choices we can make and very simple, achievable actions we can take to shift us towards becoming technology leaders (and in the process improve our profitability). These actions don't need to cost millions of dollars – many of the examples I use in this book cost a few hundred dollars or less. They aren't complex and difficult to implement (later on you will see I strongly suggest you avoid doing projects like that), and you don't need to hire a whole bunch of new staff to make them happen (because the people you need are already amongst you).

Sounds too good to be true? Welcome to the power of digital technology.

the disruption spectrum

Before we get started it is probably a good idea to find out where your organisation is starting from.

Up until now most organisations, especially small and medium sized ones, have taken a fairly ad hoc approach to their use of digital technology. The mentality has been more one of 'keeping the lights on' rather than pursuing digital innovation...and to be fair, this approach has served many organisations well over the last couple of decades. When digital technologies were slower, more expensive and less reliable than they are today, an ad hoc approach was generally good enough.

But over the last decade information technology has changed from a cost to be managed to a source of competitive advantage. Potential opportunities are now so big that an ad hoc approach is no longer adequate. The 'keep the lights on' mentality means that although systems continue to be upgraded when 'required', there are few mechanisms in place to identify and implement *new* digital opportunities.

As discussed in the previous chapter, more recently we have seen some organisations start to take a more proactive approach to digital technology. They are becoming more innovative, not just in the products and services they sell, but in the underlying approach to how they operate[4]. In doing so, these digital leaders are not just driving change, they are threatening the survival of others in their industry.

So, what type of organisation do you work in?

The following 'spectrum' is a simple way of understanding the different approaches organisations take to digital technology. More importantly, it can serve as a starting point for a robust conversation in your organisation about where you are currently at and where you would like to be.

---

[4] Interestingly there seems to be little correlation between the two, from my experience with organisations who sell technology products and services. We might accept a need for innovation in our products, but not in the way we operate.

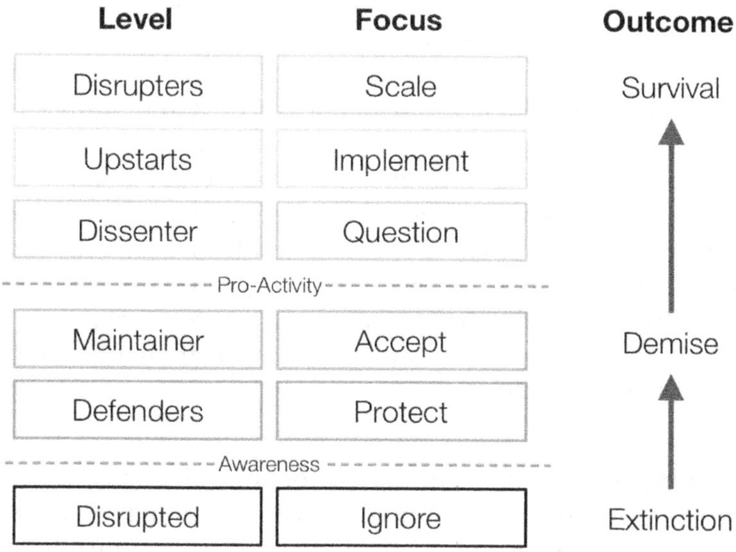

Illustration 1 - The disruption spectrum

The disruption spectrum is defined at one end by the Disrupters, the digital leaders driving change, and at the other end by the Disrupted, the organisations who feel the impacts of technology but are often ignorant of the cause. In between there are four other approaches: Defenders, Maintainers, Dissenters and Upstarts. Let's start at the bottom and work our way up.

The organisations that are **Disrupted** often don't see it coming, or if they do, they see it too late. They may identify a competitor who is out-competing them, but not the reason why. Their sometimes limited beliefs about and understanding of technology mean they are unable to appreciate its importance and are likely to blame other factors.

Above the Disrupted there are the **Defenders**. The Defenders may have a level of awareness that allows them to identify potential threats but they make a conscious decision to protect old technologies and old ways of doing things from new ones. In doing so, they apply scarce resources towards protecting the status quo instead of developing innovative practices. Somewhat ironically, Defenders often become Defenders because of their success, and they may continue to achieve short term financial gains by protecting their 'cash cows'. Unfortunately, in the medium to long term they are likely to face abrupt and rapid disruption if they can't then make change quickly.

Maintainers have a better grasp of digital technology and may even see it as beneficial. What they don't necessarily appreciate is the speed of its change and the magnitude of its impact. Maintainers don't generally see technology as being core to their business, and as a result they still view technology as a cost rather than a source of value. This results in them continuing to take an ad hoc approach to technology upgrades and new digital projects, and more often than not these projects focus on satisfying pre-existing needs rather than identifying new opportunities.

Dissenters represent a transition between old ways of doing things and new ways. The most significant shift is that they move from being ad hoc in their approach to being proactive. Rather than waiting for new ideas to come to them and relying on traditional sources of advice (which often represent vested interests), they actively seek out alternatives[5]. They may not yet understand how their organisation will look and operate in this new world, but they know that it won't look like the status quo. Dissenters are primarily driven by curiosity and a belief that there is always an opportunity to do things faster, cheaper or simpler than they have been done before.

In some ways the **Upstarts** are just like the Dissenters but with a bit more confidence. Rather than seeking out the answers they are creating them. Without wanting to sound corny, it is like the quote from Alan Kay: "The best way to predict the future is to invent it." Whereas for Dissenters digital technology is often a thought exercise, for Upstarts it's a basis for experimentation. As a result they generally operate at a level of productivity and effectiveness that other organisations could only dream of[6].

At the top of the spectrum are the **Disrupters,** who are changing the way we do business. The Disrupters are often just Upstarts with more experience and better resourcing. For Upstarts, digital experiments are often small scale and isolated (at least initially). Disrupters' previous success means 'disrupting' has become an integral part of how the organisation operates. As their knowledge and experience builds they take on bigger projects and the opportunities shift from being predominantly operational to more strategic in nature.

---

[5] If you bought this book and have read this much without muttering 'bullshit' under your breath, there is a good chance that you are at least this far up the spectrum.

[6] If you or someone else has already been appointed the digital champion in your organisation, then there is a good chance that you are at least here.

In a virtuous circle, the Disrupters' willingness to adopt new technologies early and learn from their application means they are in a better position to identify the next generation of digital opportunities more quickly than competitors. This allows them to invest early, build the knowledge and obtain further advantage.

Now take a moment to reflect.

Where does your organisation currently fit, and where does it need to be?

Are you *comfortable* with your current approach or does something (or someone) need a shake up?

I thought so.

The disruptive change we are seeing in the world right now is very, very real. In the words of Carl Sagan, "Extinction is the rule. Survival is the exception", and we need to see that in an increasingly dynamic world the approach we take to digital technology may determine our organisation's very survival.

To survive in a world of digital disruption it will take more than just developing a 'healthy respect' for the approach of Dissenters, Upstarts and Disrupters. These approaches will need to be actively embraced.

the changing face of IT

You could be wondering by now how the approach I am advocating differs from what you are already doing. The best way of explaining this is to look at how digital technology has evolved and how our approach to managing it needs to evolve with it.

The digital revolution we are now experiencing has its roots in the early days of computing. In those early days things were quite different than they are now. In the mainframe era of the 1960s and 70s early computers were much less user friendly and much more expensive than they are today. In those early days computers could only be operated by computer scientists and there were only a handful of projects that could possibly justify the necessary investment. Quite literally, unless you were trying to put a man on the moon there was little need for a computer.

Of course this changed significantly in the increasingly poorly named PC era[7] of the 1980s and 1990s. During this time organisations set up their own IT departments which served to bring technology and business closer together. During this time we found thousands of new potential projects in the cracks between the moon sized ones. This included digitising our accounting systems, and the introduction of word processing, spreadsheets and email. Each one of these projects had a significant impact on organisational productivity, to the point that our modern organisations couldn't function effectively without them.

By the early 2000s most of the PC sized projects had been identified and implemented and in many organisations the focus of IT strategy shifted from pursuing new innovations to reducing costs. In many cases IT was seen as a commodity and 'non-core' to the operation of the organisation and subsequently it was outsourced, often to the lowest bidder. Statistics from Deloitte suggest that for large organisations 40-60% of IT functions are outsourced[8], and for smaller organisations it is likely to be significantly higher.

---

[7] Poorly named because it turns out that the PC, or personal computer, wasn't really that personal compared to the mobile and wearable devices that we use today. I think we would all agree that our smartphones are far more personal than our old desktops ever were.

[8] Deloitte 2014 Global Outsourcing and Insourcing Survey http://www2.deloitte.com/content/dam/Deloitte/us/Documents/strategy/us-2014-global-outsourcing-insourcing-survey-report-123114.pdf

Around the same time as the PC buzz wore off, we started to see another significant technology shift. Just as the power of the mainframe was scaled down to the size of the PC, the power of the PC has now been scaled down to the size of a phone (and in the very near future it will scale down to the size of a button). Driven by innovations in mobile, social and cloud technology, thousands of new opportunities are now springing up in the gaps between the PC sized projects we have previously implemented.

Sadly, because many IT departments are operated with a cost focus and limited resourcing, many of these new opportunities are not being proactively identified and implemented. Part of the reason is the divide that exists in organisations between the IT function and the rest of the business. On the organisational chart (as shown in Illustration 2), the IT department is most often depicted in their own separate sealed box off to the side, as a subset of corporate or shared services, quite separate from the organisation operational functions IT is meant to be supporting.

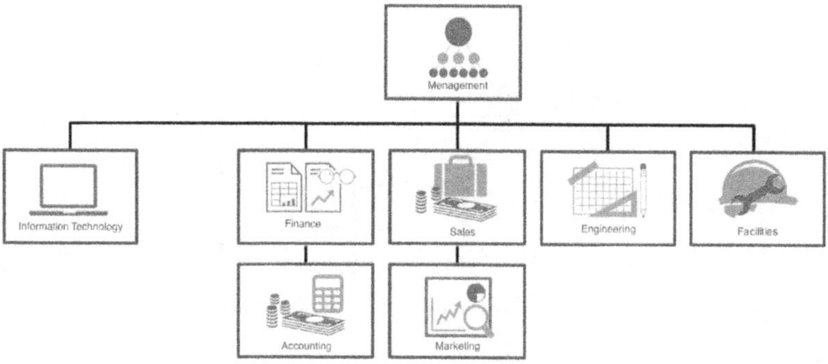

Illustration 2 - IT as we depict it

But when it comes to the allocation of accommodation, the reality is often much worse, with IT commonly housed on its own separate floor, safely removed from the rest of the organisation (all the better if some basement space is available). As a result there are often multiple floors on top of a multitude of policies, forms and budgetary constraints that stop IT effectively supporting the people they are meant to serve. And if the IT function has been outsourced then the geographical divide could be much bigger still. IT could be housed in another building, another state, or an entirely different country (with the people working in it often serving a different master).

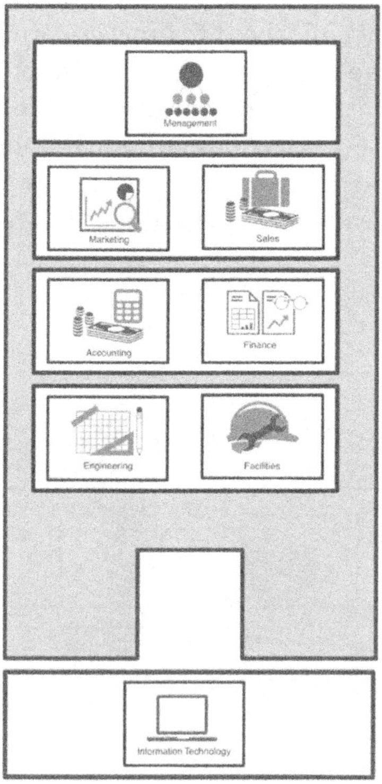

Illustration 3 - IT as it is in reality

To tap into this next generation of digital opportunity the primary focus of IT needs to shift from stability, security and cost to enabling innovation and value. This is not to say that stability, security and cost are not important objectives; in a world of digital business they are incredibly important. But if they consistently get in the way of an organisation's ability to identify and implement digital projects that provide a competitive advantage and deliver bottom line returns, then these objectives in isolation might not be serving the organisation.

The answer is not to trade out our current approach to IT for something else (though in some organisations that may be a good strategy), but rather to implement an additional mechanism to bridge the divide between the business and technology sides of the business. Organisations need a way of working directly with their people, the end users, to enable a new generation of digital opportunities, and as luck would have it, this is what we will be exploring in more depth in this book.

Many of these opportunities may appear to be relatively small and insignificant compared to the types of projects that IT departments have traditionally delivered, but when you add them together they provide a substantial return, and often with lower levels of risk than have been undertaken in the past.

IT departments are operating in a marketplace

I have no doubt that the digital champions approach represents a significant shift for some people working in IT. With each new platform comes new risks. Integration risks, security risks and stability risks. In addition there are expectations of support being provided, which requires resourcing and money, plus the cost of the platform, app or software being used.

I get it. But I don't accept it.

IT departments once had the luxury of operating in a monopoly market (being the sole service provider to their organisation) but this is no longer the case. The growth of web based software as a service platforms and increasingly tech savvy users means that IT is now competing in an open market against other service providers. Service providers who somewhat ironically know much more about what the organisation's people need than their own IT department.

If IT departments don't work with their people to provide them the digital tools that they want, there is very little stopping people from helping themselves. The most likely outcome is that they just won't tell the IT department they've done it (because IT might just go and shut it down).

The impact of this is that the organisation faces the same or potentially higher levels of risk, the same challenges of interoperability and much the same, if not higher costs. By being excluded from the conversation, IT can't provide advice on the best options and can't help when things go wrong.

To be relevant in the 21st century, and even to fulfil its existing mandate for security, IT departments will need to develop a more customer focused approach to how they deliver their services.

connecting the dots between people, work and technology

Digital technology projects have a bad reputation for running over budget and underperforming. In fact, research by McKinsey suggests that, on average, large IT projects run 45% over budget and 7% over time, and deliver 56% less value than predicted[9].

When McKinsey looked at the underlying cause, they found the number one cause of these failures was that projects did not have a clear business case. Fundamentally,

---

[9] Delivering large-scale IT projects on time, on budget, and on value http://www.mckinsey.com/business-functions/business-technology/our-insights/delivering-large-scale-it-projects-on-time-on-budget-and-on-value

this lack of alignment between the business and technology means that:

1. The people who understand the business rarely understand the technology

2. The people who understand the technology rarely understand the business.

This challenge is becoming increasingly apparent and many large organisations are now taking steps to address it. The number of Chief Digital Officers (CDOs) in the world is currently growing at 200% per year. CDOs are taking a focus on digital to the very top of organisations and this extraordinary growth reflects an increasingly proactive approach to digital technology by large organisations.

Although CDOs can fill a number of different functions they are often tasked with facilitating the adoption of digital tools by the organisation. They understand the business need first and the technology a close second. They are advocates of the organisation's people but can also communicate effectively across stakeholder groups.

In smaller organisations, budgetary and resourcing constraints mean a full time CDO is unrealistic. For smaller organisations this is the role of a digital champion. Someone who can provide a similar set of skills as a CDO but on a smaller scale and a part time basis.

Whereas the IT department might be considered the builder of digital solutions, the digital champion is more like the architect. Through an intimate knowledge of how the technology will be used the digital champion works with the CEO and the organisation's people to determine what is possible and define practical solutions to meet their needs.

As the illustration below shows, it is the ability to incorporate and address the competing needs of stakeholders that defines the digital champion's role. Fundamentally there are four key stakeholders that need to be considered, and none of these are doing an adequate job of identifying and managing digital projects individually. This idea is depicted in the diagram below.

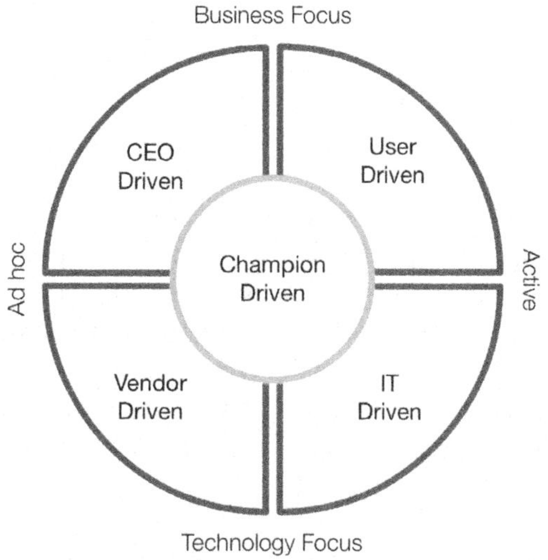

Illustration 4 - The key stakeholders of the digital champion

The first is the executive. Although they understand the business need and are integral to the resourcing of digital projects, their commitments elsewhere in the organisation mean their time is limited. They are a key stakeholder whose ultimate concern is in ensuring that projects deliver a suitable return on investment.

The second is the IT department. Once again, an important stakeholder when it comes to stability, security and integration but one that has traditionally failed to truly understand and address the business need. IT has the time and the commitment to enable the business but its traditional structure and resourcing means that it often lacks the ability to find opportunities proactively.

The third stakeholder group is vendors. They are also integral to the process but in and of themselves are generally biased towards finding problems that fit their solution. Also, as pointed out in the foreword, I'm a vendor and as an outsider I can never truly understand your organisation (and what it needs) like someone on the inside. Ideally what a vendor is looking for is an informed customer that can help guide the engagement and build internal support for change.

The last group is the end users, or our people. This is the most important group of all. These are the people with the problems they need help with. Our people are increasingly tech savvy and many of them are already implementing their own solutions on the sly. The problem is that they lack a structure and approach to doing this safely and have no clear mechanism for engaging with IT in a way that doesn't result in their toys being confiscated.

If only there was someone who could understand and balance these varying needs. The executive's need for return on investment, our people's need for usefulness and usability, IT's need for stability and security and even the vendor's need for structured implementation and long term revenue.

Enter the digital champion. Our digital champions need to communicate and influence with business side stakeholders as well as the technology side ones. They need to get the business first and the technology a close second. They will also need the resourcing to identify new opportunities and a method to ensure they are assessed and implemented effectively.

The time has come for digital champions.

HELLO DIGITAL CHAMPION

so what exactly is a digital champion?

So far we have a rather nice slogan, 'to connect the dots between people, work and technology', and a few stakeholders, but what is it that a digital champion actually does?

A digital champion's primary objective is to help solve business problems using digital technology. First and foremost this is a business improvement role and although a digital champion may have an IT background (but this is not necessary as we will discuss later) their approach should always serve the business first and the technology second.

Successful digital champions generally share three key characteristics. The first of these is generosity. Digital champions need to love listening and love helping people. A good digital champion needs to have empathy for people who don't necessarily 'get' the technology and are often quite fearful of it (undoubtedly the digital champion will also be called upon to do a little 'help desk' work every now and then). They have to take the time to engage directly with the people using products (face to face rather than SMS) to understand the problems behind the problems that they are dealing with. A good digital champion will never forget that their role is primarily about solving people's problems, not implementing digital solutions.

The second characteristic is curiosity. Digital champions should be constantly testing out and trying new things. The rate of change in the digital space means that there is an ongoing battle to keep up with the latest developments[10] and so curiosity, a strong desire to know and learn, is an incredibly important part of staying relevant. One way of looking at it is that generosity is what drives the digital champion to understand the problem, curiosity is what helps them identify potential solutions. Either one without the other won't deliver value back to the organisation. In fact, curiosity without generosity is likely to be a constant form of distraction.

The third characteristic of a digital champion is laziness.

Yes, laziness.

I once read a great blog post that the best developers are lazy[11]. They will program in a way that minimises the amount of code they need to write, which subsequently

---

[10] A good digital champion will enlist their army of users to help identify and test new ideas but it is unlikely that the digital champion will ever stop being curious.

[11] https://medium.com/@vinodkumaar/the-lazy-programmer-27ae8abdd8ff#.gx0pjgcb0

minimises the amount of code that a computer needs to run. The outcome is a reduction in the amount of computing resources and time needed to run the program.

I think at this point it is important to distinguish between laziness and disinterest. Disinterest is when I don't want to do the work, while laziness is when I want to do as little work as possible but still deliver the outcome. Someone who is truly lazy will do more work upfront to do less work later on. In effect the principle of laziness is the basis of all business improvement activities; it is the idea of working smarter, not harder.

Digital champions should always be looking for the laziest projects, the ones that require the least amount of effort to implement but still generate reasonable returns (see the chapter 'Rabbits and Rhinos' for more on this). They should also be capable of understanding and engaging the business based on the return on effort, or how much time needs to be invested to save much more time later on. This return on effort is a key part of the business case that digital champions will develop as they investigate their projects.

the digital champion's approach

Alright, we now have a rather nice slogan, a bunch of stakeholders and three common characteristics of a digital champion (generosity, curiosity and laziness)...but still no actions. Well, you will be pleased to know that almost the entire rest of this book is devoted to what a digital champion actually does. But all those subsequent pages can be summarised as three key activities.

A digital champion identifies possibilities, investigates opportunities, and implements projects. As you can see in the illustration below there is a connection between the characteristics of a digital champion outlined in the previous chapter and these three activities. The ability to identify suitable projects requires a combination of laziness to identify the problem and curiosity to identify potential solutions. Investigating opportunities requires curiosity, too, but also the generosity to work with stakeholders and help meet their diverse objectives. Finally implementing projects is best done with both generosity and serious thought on how it can be done with the minimum amount of disruption and effort.

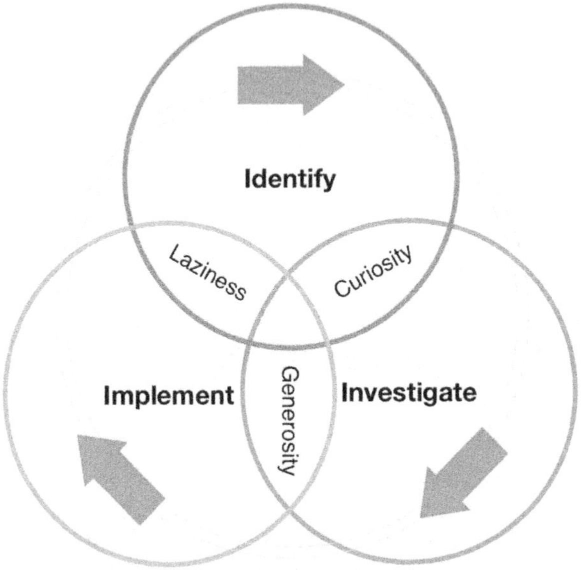

Illustration 5 - The activities of a digital champion

Is that it? Pretty much. In the subsequent chapters each of these activities will be expanded on in more detail but for now it is useful to think of them as a funnelling and filtering process for possible digital projects your organisation could undertake. The widest part of the funnel contains the possibilities, which are then narrowed down to opportunities and finally reduced to a smaller number of projects[12].

At the identifying stage our options are both numerous and fuzzy. They may consist of a common business problem that is currently lacking a solution, or even a solution (an app, platform or piece of software) looking for a problem to fix.

As we gather more information some of these are prioritised and more effort is put into their investigation and development. These are the genuine opportunities and in this stage the digital champion's role is to try and assess their suitability and potential return. Some of these will turn out to be unsuitable and others will be postponed but the rest will move towards implementation.

At the implementation stage the funnel gets narrower again and it is still possible for projects to be withdrawn or postponed. Even at this point there will still be the *known unknowns* and the *unknown unknowns*, as Donald Rumsfeld so eloquently put it. The known unknowns are the potentially project breaking assumptions that need to be tested to prove the viability of the project. The unknown unknowns are the potentially project breaking assumptions that also need to be tested...but we just don't know they exist yet.

---

[12] In the above illustration this is represented as a Venn diagram because at any point in time a digital champion is likely to be doing these activities concurrently, even though for a particular project they follow a specific order.

These three activities are sequential for any one project, though depending on the size of the projects undertaken a digital champion may manage multiple projects concurrently. As they are doing implementation on one project they are likely to be investigating another, and while working with users to investigate or implement projects there is always the opportunity to uncover new unknown problems and new possibilities.

Ideally, a digital champion should always have a runway full of potential projects ready for take off. As they are finishing up one they will be already starting on another. As pointed out earlier the average size of each project that a digital champion manages might be small, which means that more projects can be managed at once.

the six phases of a digital champion

Individuals coming into the role of the digital champion normally come from either one of two backgrounds: the business side of the business, or the IT side of the business. My background was in the business side of the business. It provides a distinct advantage when engaging management and an organisation's people but gives me little sway with IT. On the other hand, some of the best digital champions I know have an IT background, and they often experience engagement challenges that are the polar opposite of mine.

I call these two groups of potential champions the Users (the business side of the business) and the Experts (the technology side of the business). They represent the first two of the six phases that make up the digital champion's journey. Although the Users and Experts are presented in a hierarchy in the illustration below, one does not necessarily have an advantage over the other[13]. Invariably they understand the business or the technology, but rarely do they understand both. The real opportunity starts to emerge when these two perspectives intersect, which is the domain of champions moving into the third phase: the Explorers.

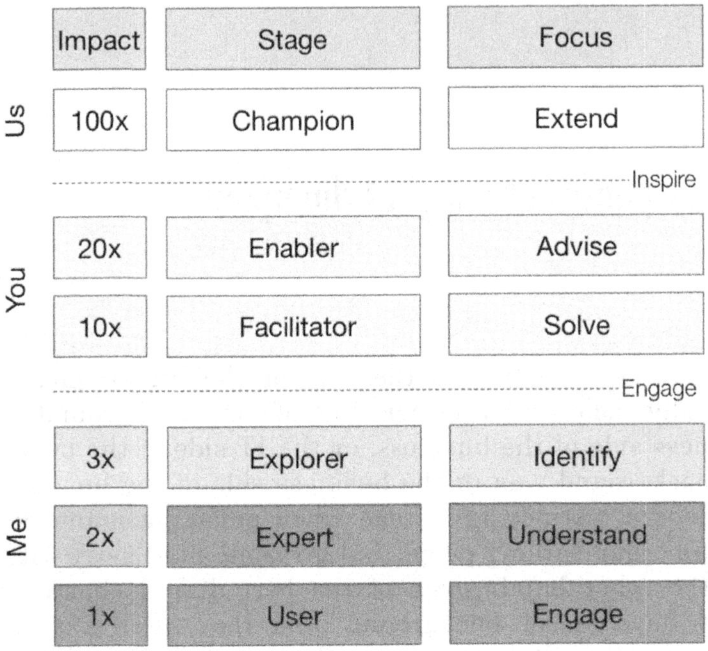

Illustration 6 - Development of a digital champion

---

13 ...though the Expert might have slightly more impact in that they can at least recommend pre-approved solutions to others.

Explorers are Users or Experts who start to look at the business opportunities of technology, or the technology opportunities of business (depending on which side of the divide they started on). Initially, the value of this is limited because it is mostly done for personal benefit rather than to help others. Users doing this on the business side of the business often keep their exploration under wraps on the belief that if IT knew it would probably be taken away.

For individuals to progress from these bottom three levels, an organisation generally needs to formalise the role of the digital champion. Without making this role explicit and providing legitimacy for digital projects, it is almost impossible for potential digital champions to start engaging with others and shift from an inward 'me' focus to an outward 'you' focus.

The first of the outward focused roles is that of the Facilitator where the emphasis is on working with others to identify and solve their problems for them (the *doing* and *managing* aspects of the digital champion's role discussed in the next chapter). As you would expect at this point, there is a significant bump in impact as new opportunities can be discussed openly which allows them to spread to others in the organisation.

As the experience of the digital champion grows, and assuming that the organisation continues to mature, the digital champion eventually shifts to an Advisor role, the last of the outwardly focused roles. Instead of identifying and solving problems for others, the digital champion can scale both the impact and influence by advising others and enabling them to do this for themselves.

The final stage requires another significant shift in focus, from 'you' to 'us'. This is when the digital champion becomes a Champion. The role becomes more strategic and the focus shifts to influence. Here, a digital champion can have the most impact and influence, but that requires a certain level of organisational maturity. More on that in a sec.

four types of projects and four different approaches

The types of projects that a digital champion works on and how they are implemented will depend a lot on the maturity of the organisation they are working for. The illustration overleaf identifies four key levels of maturity, defined in part by the types of digital projects organisations are willing to pursue, from Conformity, to Productivity, then Quality, and finally to Positioning.

So although a digital champion might identify some awesome projects to work on that could deliver significant value back to their organisation, depending on the level of maturity, the organisation may not be ready for them. That's OK, you can always just put those projects in the back pocket and whip them out again later.

Illustration 7 - Four levels of organisational maturity

The first stage shown in the illustration above, and the existing state for many small and medium sized organisations, is conformity. Conformity is what happens when IT departments are focused primarily on stability, security and cost. The principle behind conformity projects is that if we can get everyone to conform to the same standardised system then it will be cheaper and easier for the IT department to manage and maintain (in general the value of conformity is measured in $1,000s). At the level of conformity the approach to engaging people is coercion, primarily because it means taking away from people other, often better, options than they could have otherwise had.

As pointed out earlier, the conformity approach is starting to fail and only the most naive would believe that their people aren't already colouring outside the lines and using 'nonapproved' digital tools. As well as this, coercion as an engagement approach is rarely effective and doesn't encourage two way dialogue; and without two way dialogue, how can you ever really find out what problems people are dealing with and what solutions they really need?

In the second stage, the objective moves from conformity to projects that improve productivity. This requires a significant shift in terms of how IT is viewed in the organisation. In the conformity stage, IT is perceived as a cost to be managed. In the productivity stage, IT needs to be seen as a driver of value.

For digital champions, productivity projects are likely to be the first types of projects that you will work on, because they are generally the easiest to justify. Productivity projects save people time. We have all heard that time equals money, and when it comes to labour we have a fairly good understanding of how much time is worth – about $52 per hour. This is based on the average Australian salary of about $70,000, plus on costs and overheads such as superannuation, management, IT and rent, equating the cost of a full time employee to around $100,000 a year. For that, you will get 40 hours of someone's time for about 48 weeks a year, which works out to be about $52 per hour.

So for every hour a productivity project saves it equates to $52 per year in benefit. If you can save someone an hour a week it is worth about $2,500 per year, and if you can save 40 people an hour a week, suddenly you have a $100,000 return to the business each and every year.

Just in case you weren't paying attention, let me say that again. In an Australian business if you can save 40 people one hour a week, you have a project worth $100,000 per year.

BAM!

The value of productivity projects are generally measured in $10,000s but as you can see $100,000+ projects are quite conceivable. With this amount of financial and time benefit at stake let's hope that we don't need to make much effort coercing people to change. At this point our engagement approach shifts from coercion to convincing. The idea here is that if we present the opportunity to the right people in the right way and show them how to achieve the benefits on offer they will hopefully join in out of pure self-interest.

At this point the digital champion's role may also start moving from doing to managing. There will still be work for the digital champion to do, but the value of the projects being undertaken will sometimes justify the appointment of outside specialist resources to help in the delivery. Coordinating and managing these resources should continue to be the digital champion's responsibility.

Of course, doing things faster (i.e. increasing productivity) in and of itself is fairly pointless. If you aren't doing the *right* things, you won't be heading in the right direction.

At the next stage organisations start to appreciate the folly of just doing things faster for the sake of it and want to do things better, and subsequently the types of projects that digital champions focus on change.

With productivity projects the objective is often to do the same process in fundamentally the same way, but faster. An example of this is trading out physical mail for email. The message will get to its destination faster, cheaper and more reliably than before but it doesn't change the underlying reason as to why the message was sent. When you shift to quality projects digital champions need to start thinking about how digital technology allows the process to be modified. For instance, if the email contained a weekly status report that needs to be sent from one department to another, perhaps the report could be auto-generated in real time and posted on a webpage so anyone who needs to can access it.

The biggest challenge with quality projects is that the value proposition is much harder to articulate, though generally it's a multiple of value when compared to productivity projects[14]. In general quality projects are valued in the $100,000s, though often that is difficult to quantify. The good news is that by this point a digital champion should have already delivered so much value to the organisation and their people through their productivity projects that they no longer need to convince people to do them.

At this point the engagement approach employed by digital champions should shift from *push* to *pull*. At both the conformity and productivity stages technology is effectively *pushed* onto the user (though at least with a semblance of two way engagement with productivity projects). This is because at the moment, the value of the technology (and maybe even the IT department more generally) is at best unproven and at worst perceived as negative. But through the delivery of simple, small projects that solve people's problems, the perception shifts. At some point the digital champion will no longer need to be the identifier of problems and the instigator of projects. People will come looking for

[14] In the above example posting a real time report on a webpage might only save an hour a week (worth about $2,500 per year) but if that up to date information results in a better decision being made, then it could be worth tens of thousands or even millions of dollars, although we would never really know.

digital solutions to the problems they have identified themselves, effectively creating a *pull* for technology in the organisation, and the approach of the digital champion shifts to that of advising and enabling.

If the organisation and its people are demanding technology solutions, then there is less need for a value proposition to justify the project and engage others. I don't mean to say that there shouldn't be a value proposition, just that how we develop and articulate the value proposition changes (more on this can be found in 'Investigating Possibilities').

Now, the digital champion will do less doing and start to play more of a support role, helping people (who are becoming increasingly tech savvy) to do the doing for themselves. At this point people are already identifying their own projects and the digital champion will help them investigate and implement their own projects. They will still help engage and manage external resources and get their hands dirty when necessary but their primary focus is on enabling others to become their own digital champion.

The last stage in organisational maturity when it comes to digital projects is positioning. Much as the value of quality projects can be a bit nebulous compared to productivity, positioning projects can be just as nebulous when compared to quality ones, though generally positioning projects are valued in $1,000,000s. The difference between quality and position projects is the difference between current and future value. Quality projects are focused on doing the stuff we do in the present better than we did before. Positioning projects are about preparing our organisation to do better work in the future.

There are two types of positioning projects. The first have a short to medium term focus and aim to position the organisation's *existing* skills and knowledge in a different way or to a different market to attract new business. The second type of project has a medium to long term focus and involves *upgrading* the organisation's skills and knowledge to position it for emerging opportunities, especially in the digital space.

Both of these projects are quite strategic in nature and the value is difficult to quantify until after the fact. As such it is likely (though not necessary) that a digital champion has already completed a number of productivity and quality projects before embarking on the positioning ones. Assuming that they have done their job well, a digital champion's approach to engagement can move from advising to inspiring. The focus is now on bringing outside influences into the organisation to guide conversations about what is possible and where the organisation wants to go.

By this point the role of the digital champion should have become very strategic in nature. There will be very little doing or managing left to do as people in the organisation will have taken on much of that work themselves. The digital champion can begin to act as an internal mentor[15], with the primary objective to extend their reach and subsequently the impact digital technology can have in the organisation.

---

[15] I love the distinction that my good friend (and mentor) Matt Church makes between coaching and mentoring. The underlying premise of coaching is that the coach doesn't need to know the answer but if they ask you enough questions you will work it out for yourself. On the other hand, a mentor has previously been there and done the thing you are setting out to do. Unlike the coach, the mentor has the answers to questions you want to ask, even if you didn't know you needed to ask them.

An organisation will not necessarily mature evenly through each of these stages, and therefore a particular level of maturity in project objective as shown in the previous model may not always align with the corresponding approach to engagement. The best way to think about it is that each level also includes all levels preceding it. When you have reached quality you might still deliver projects with a productivity focus, or even seek out opportunities to push conformity[16].

The only real limitation is trying to operate and pursue projects at a level higher than what you organisation is currently comfortable with. For instance it might be difficult to convince your organisation to invest heavily in digital skills development when you haven't shown the productivity benefits of digital tools and there is a general belief amongst management that IT still belongs in the basement.

The real maturity comes when an organisation has the full suite of options available to it and the digital champion has the ability to identify the right combination of approach and role for the projects they are working on.

---

[16] At the end of the day conformity projects still have a place, especially if there is an opportunity to pursue economies of scale in a particular area of your digital ecosystem. The problem most organisations face is that the focus on conformity is at the exclusion of other projects and therefore they don't pursue innovative opportunities

the problem with meetings

The best opportunity for digital champions lies in the areas where people feel there is a lot of time wasted (because they are generally correct), and a great starting point for finding productivity opportunities is meetings.

If you look at the impact of digital technology on meetings over the last couple of decades it has been somewhat underwhelming and almost exclusively productivity focused. The introduction of email and shared calendars has meant that we can arrange and prepare for meetings faster and easier than ever.

This has certainly been a productivity boost but if you look at the problems that people currently have with meetings they are largely the same problems that we had 20 years ago. Are the right people at the meeting? Are the right items on the agenda? Have people completed the actions and commitments they made at the last meeting? And ultimately, is this meeting even necessary?

So what could we do, apart from having another meeting, to figure that out?

If we shift our focus from productivity to quality, we can start to understand how digital tools could be used to make meetings better. Perhaps we could use a collaborative task management app such as Wunderlist or Todoist to track actions so everyone has visibility (and therefore accountability) for the meeting actions. Maybe we could socialise the agenda using an enterprise social tool such as Yammer to make sure the right people are there and the right items are on the agenda. And possibly, if we were to do both those things we would find out the meeting was entirely unnecessary.

developing the champion's champion

The rapid growth in the number of Chief Digital Officers around the world hasn't happened by accident. It is a direct result of the CEO and executive identifying a need to approach IT differently and then resourcing it. In much the same way, your role as a digital champion is going to require the direct intervention of key decision makers.

The number one reason for this is you will require resourcing. The biggest challenge new digital champions face is having the time to complete their projects, even when there is a substantial return on investment that has been identified. The thorough identification, investigation and implementation of digital opportunities will take time and effort and the implementation of new solutions will not be without cost. A lack of funding for the projects that the digital champion identifies will undermine the role.

Secondly, the support of the executive brings legitimacy. There are many potential digital champions who play a similar role in an unofficial capacity, identifying and testing digital tools for their own use and often helping people out with their problems, but without the explicit support of the executive this will always remain an ad hoc and unstructured approach.

Finally, support from the executive provides influence when required. Over the medium to long term the influence of the digital champion should be a result of the outcomes they generate. But in the short term the support of the CEO or executive may be required to build support amongst stakeholders for both the role of the digital champion and the projects they are implementing.

So how do we develop our internal champions?

Developing internal support for such a new role is not without its challenges. As a result mechanisms to create accountability and measure success are paramount.

The starting point for developing internal support is covered in the previous chapters, especially those outlining the changing nature of IT and the risks of digital disruption. This is the 'burning platform' that is key to trying something different. Make the time to discuss these with your manager or CEO or even offer to run a lunch and learn session on the opportunities of digital technology.

As part of this identify at least one opportunity and calculate the potential return (you will find out how to do this in subsequent chapters). The return needs to be significant enough to get your bosses' attention and address a problem that matters to them. Now is the time to float the idea of the digital champion.

For most small businesses finding and hiring an external digital champion will be a difficult, and possibly even unadvisable proposition. Difficult because people with the digital champions skill set are rare and in demand, and unadvisable because the best digital champions are ones that know the business intimately. As a result, most will be left with little choice but to develop digital champions such as yourselves.

If they buy into this proposition the next objective is for you to de-risk the implementation. This can be done through a clear commitment, open accountability and ongoing feedback. The commitment is a trade of x for y. In the Digital Champions Club it looks something like '20% of someone's time in exchange for $100,000 in agreed value each year'. Without a clear agreement of what success looks like and the commitment that each of you are making, the idea of a digital champion will always struggle to get traction.

Accountability mechanisms should be important to both yourself and to others. Personally I'm not great at being accountable to myself but I'm pretty awesome at being accountable to others (I don't like letting other people down). For instance I make one of my team accountable for my blog posts because that way if I don't write them I am actually letting them down, not me. Think about who you will let down if you don't meet your obligations as a digital champion and how you can make information open and accountable to others. Maybe you agree to have two projects investigated and actioned in the first three months as a 'trial period' before your role as a digital champion is confirmed.

Lastly, make sure that part of your arrangement provides you both the opportunity to receive and give regular feedback. It might be that initially you get together once a month to review the projects you are working on and the progress you have made. Regular feedback will not only drive you to get more done and help your internal champion better understand the digital champion's role, it will also provide an opportunity for you to identify strategically significant projects. Identifying projects, especially strategically significant ones, will be instrumental in building your reputation.

IDENTIFY POSSIBILITIES

rabbits and rhinos

Every project, whether it is a digital one or not, has some unique combination of complexity and payoff. When choosing which projects to undertake first it makes sense that we would want to focus on the ones with the lowest complexity and the biggest payoff.

The matrix shown in the illustration below looks at the four different types of projects we can pursue based on a simple breakdown of effort (low effort or high effort) and return (high return or low return). In a way, this is a matrix that helps us map the return on investment of our projects. Projects in the top right (the Dodos) have a high return on the effort invested and projects in the bottom left (the Dogs) have a low return. Projects in the top left (the Rhinos) and in the bottom right (the Rabbits) have a similar return on effort, but as we will see, they are very different types of projects.

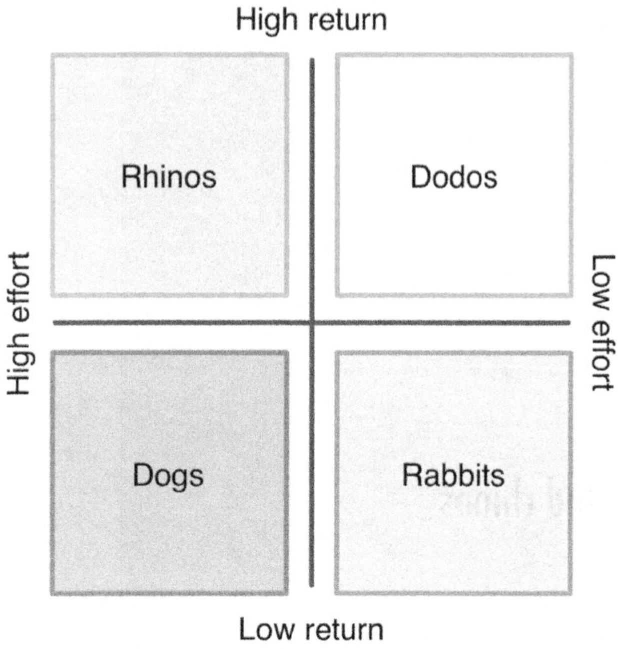

Illustration 8 - Four types of digital projects

Of course we would expect that most of the simple, high value projects have already been done. They are called the Dodos because they are largely extinct, though the speed new technology develops at and the changing needs of users means that every now and then a new simple, high value project pops up, but you can be fairly sure that it will disappear and become a Dodo almost as fast.

As pointed out above, at the other end of the spectrum we have the Dogs. These are the projects that are complex and difficult to implement and have a limited payoff. Given the amount of digital opportunity out there it would seem unlikely that we would want to pursue these projects, but there is clear evidence to suggest that we sometimes do. The most likely reason for this is that we didn't initially realise they were Dogs, we didn't truly appreciate how difficult they would be and we may have assumed at the project outset that the payoffs would be higher.

And then in the middle we have the Rhinos and the Rabbits. The Rhinos are complex projects with a big payoff and the Rabbits are the small projects with a much more modest payoff. While professing to have almost zero experience in delivering large scale digital projects and without wishing to question the long term value that they are capable of producing, I think that Rhinos are not necessarily the right focus if you are a digital champion[17].

Rhinos might look good from afar but they are often far from good. If you were a hunter in Africa and looked at a rhino from a distance you might think it was a fairly worthwhile target. As it sits there grazing it looks fairly meaty and relatively slow, and if you could catch it and kill it, you and your tribe would eat well for weeks. You would be a hero. It is only once you get closer to the animal that you realise that it has armour like plates for skin, has a massive horn on its head and over a short distance runs faster than Usain Bolt.

Rhino projects are much the same. Hidden within the complexity of these projects are unknown risks, blown out

---

[17] One of the big Rhino sized projects being pursued globally by a number of organisations is autonomous vehicles. Delivering on this involves substantial risk, will cost billions of dollars and take many years...but the potential payoff is extraordinary. The complexity of delivering on this means that for some organisations the Rhino may turn out to be a Dog and it wouldn't surprise me if some of the large auto companies fold before they can deliver.

timelines and missed assumptions that make the delivery of them much harder and often undermine the original value proposition. So although delivering big projects can be great for building your name and reputation, it often doesn't take much to go wrong to turn a Rhino into a Dog[18].

Which leaves us with the Rabbits. As you can imagine, hunting rabbits is a whole bunch easier than hunting rhinos. They certainly won't feed the tribe for a week or two at a time but they are so numerous and so easy to catch that everyone will still end up with a full belly. Rabbits are the small projects with reasonable returns. They are already numerous in nature but with the ongoing evolution of our digital tools we can expect many more of them to emerge in the future...in effect they are breeding like rabbits!

The distinction between Rhinos, Rabbits, Dodos and Dogs is relative, not absolute. Some potential projects are simpler and some projects are more complex, some projects are extremely valuable and others are less so. Plotting your potential projects on the matrix shown above is a great way for determining which projects you should focus on first. Ideally you would want to do this with input from your internal sponsor. Their feedback on which projects are strategically significant might materially impact the perceived value and subsequently which projects you choose to implement first.

It should be noted that the projects I advocate digital champions pursue are not the ones that have been

---

[18] We can pursue Rhino sized projects, but we shouldn't start with them. Once we have built our knowledge, expertise and organisational support by delivering simple projects, we can then move onto bigger ones.

traditionally targeted by IT departments. This is because the need for these projects wasn't immediately obvious and they didn't fit the profile (scale, value, complexity) of what IT departments (often with extremely limited resources) thought they should be pursuing.

The good news is this has created a fertile ground for digital champions to operate in. All you need to do is find yourself a list of potential projects.

are dodos really dead?

Whether we classify a project as a Rhino, a Rabbit, a Dog or a Dodo is subjective, relative to the other projects we have identified. But there will be some projects that have a great return, limited effort and, frankly, should have been rolled out years ago.

One example of a Dodo project that many organisations seem to have missed is showing people how to turn off their email notifications on their laptops and smart phones. This is an incredibly easy project to roll out and has an immediate impact on both productivity and the quality of our work.

Research suggests that when we get distracted it takes on average 23 minutes and 15 seconds to get back on task, and one of the most common distractions is notification messages. Most likely, before the 23 minutes and 15 seconds is up, we receive another notification and subsequently operate in a state that technology commentator Linda Stone has called continuous partial attention. In this state both the quality of our work and our personal wellbeing suffers (as we are operating with higher levels of stress), which is only made worse when we then feel we need to work outside of normal hours to catch up on the work we were distracted from.

If you can provide people with a 30 minute lunch time training session and show them a simple approach as to which apps should and shouldn't be able to disrupt your work...well, they will love you forever!

but wait, how do I find suitable projects?

So we have a methodology for determining which projects we should be working on, now we just need some projects. As pointed out in the previous chapter, the types of projects we are looking for are not the types of projects that IT departments normally pursue. As a result they may not always be obvious, even when they are in plain sight.

There are a number of strategies that we can use to identify projects but there is one strategy that trumps all others – and I mean ALL of them. To *not* pursue this strategy will dramatically impact your effectiveness as a digital champion. Now that I think about it, you can't call yourself a digital champion unless you are willing to pursue this strategy, and pursue it often.

The number one strategy for identifying opportunities is, wait for it...

Talk to the people who use the technology.

I know! Revolutionary stuff. The reason why this is such a fertile ground is that most of the time people who work in IT *don't* talk to their people, or when they do it is because something is broken or not working properly. Normally it is a fairly unenjoyable experience for both parties and not one that lends itself to a discussion on what new digital opportunities we could pursue.

Without talking to your people you can never possibly know what they need, and without knowing what they need, you cannot possibly find a meaningful solution, and without a solution (that they actually want) you won't be able to create wondrous amounts of value together.

Later on we will discuss this in more detail as talking to our people is also the number one strategy for investigating possible projects. In fact it is during these investigative conversations you often have the best opportunity ever to identify other possibilities by asking a few probing questions. Initially we might not have this type of positive relationship with our people and as such asking probing questions might be a little awkward. When we first start out on our digital champion journey it is therefore good to have another approach up our sleeve.

One way of kicking off a great conversation about digital possibilities with your people is to undertake a value stream mapping (VSM) process with some of them. By mapping out the process of how information flows between people in an organisation, you will more than likely identify places where it is being done really badly. Examples might be people using Excel spreadsheets for managing projects and tasks, or Excel as a CRM (Customer Relationship Management) system...or Excel for just about anything else. Other common opportunities I have found include tracking expenses and compiling credit card reports and sharing documents using cloud tools to avoid version control issues.

Running a VSM exercise has *never* in my experience failed to both identify potential projects that users want straight away *and* deliver huge amounts of immediate value back to the participants. Most organisational processes have been altered, changed and bolted onto over many years and even those embedded in the process don't really know exactly what is involved any more. Just by taking the time to talk it out helps to clarify a whole bunch of non-technology related opportunities to improve how they work together.

Now, if you aren't talking to your people yet (and why not, I ask you? That piece of advice is at least two paragraphs old) and you don't feel you have the knowledge, skills or confidence to run a VSM exercise there is one final option available to you: look for the paper.

For a long time our digital tools got in the way of our analogue systems. The way information moved both through organisations and between them was paper based. In those times capturing, storing and sharing information in digital format was often problematic. The end result was that people would usually print stuff out, stuff it in an envelope and post it off to people.

Now we have the opposite problem. Now our analogue, or paper based information, is getting in the way of our digital systems. The standard for sharing information between people is now typed text (more email anyone?). Anything that doesn't start out as digital, or started as digital and then got printed, not only comes packaged up with a whole lot of bad tree karma, it also represents a great opportunity to improve organisational efficiency and effectiveness.

a little bit about value stream mapping

Value stream mapping (or VSM) is a Lean business improvement tool that involves mapping out a process from the end to the beginning. To do it effectively you need all the main contributors to the process in the room together. At least part of the value is created when individuals find out what other people do in the process and how their actions impact others.

A complete VSM is a fairly time consuming process but you can run a simplified version for the purpose of identifying potential digital projects. Whereas VSM is most commonly used for mapping out physical processes, this simplified version has been designed for use with information based processes. For the exercise you will need at least three different coloured post-it notes, a bunch of markers, and a whole lot of wall space (tape paper to the wall if you want to take the map down to save or work on over multiple sessions).

To start you need to agree on the process that you want to map and establish the end point – the end point is normally seen as the point of value creation – and an idea of the starting point which you are working backwards to (at this point the starting point can be a little fluid). Because we often deal with processes built on top of other processes that bleed into yet other processes, it can sometimes be a little time consuming to get agreement on this. Unfortunately, until this is agreed you don't really have a process to work on.

Next, as shown in the illustration below, starting at the end map out three elements for each stage in the process. The first element, done on one coloured post-it note, is the task or activity that was completed. The second element, done on a different coloured post-it note, is the person who does it. On the third post-it note, jot down the information type and/or platform used in this stage of the process.

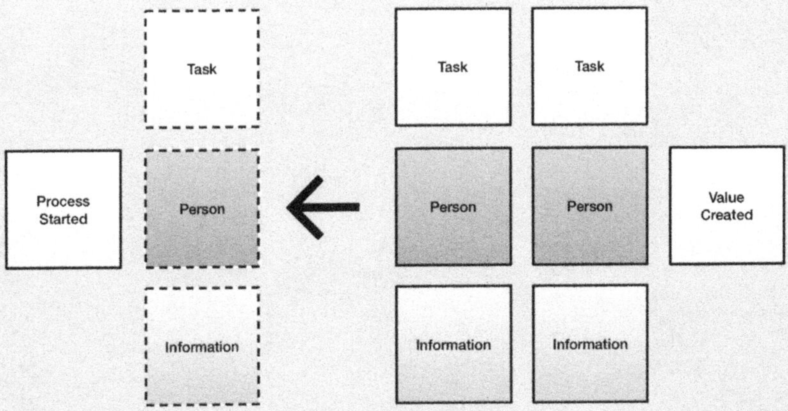

Illustration 9 - How to create your value stream map

As you work backwards through the process being mapped look for points where information flows poorly (this might include information being lost, double handling, rework or information flow being just plain slow). In the gaps between the post-it notes, flag the problem, and if you are aware of a platform or tool that might help fix it then flag that as well.

Once you have completed mapping the process you should have at least half a dozen opportunities that you have identified. The next step is to engage the people who participated in the VSM process as supporters. Later on, when it comes to investigating and implementing projects, supporters will be invaluable. To get them on board it is always useful to be able to show them what the alternative might look like. If you can demonstrate an app or platform that is a potential solution you can both build support and get immediate feedback. It is incredibly important to start creating demand for the technology you are looking to implement, and the easiest way to do this is if your people can see what's in it for them.

INVESTIGATE OPPORTUNITIES

the story of your project

I am a huge fan of Patricia McMillan's book, *Make IT Matter*. Patricia is a thought leader helping people in technology tell their story. This might be a startup trying to articulate the value of their product or an IT department trying to convey their strategy to people who don't speak geek.

In her book Patricia says that IT departments need to have three stories. The first story is their story, why they do what they do. The second story is the story for the people, why someone would want to use the product or service they offer. The third story is for the enablers (the people who make the project happen), how the organisation will benefit from the project.

As we investigate our projects as digital champions we want to build out the second two stories, the story for our people and the story for our enablers. I agree with Patricia that the first story is important, and I strongly suggest you work on it, but for now we will focus on the other two.

The idea of creating a story serves a number of purposes. Firstly and most importantly it is a vehicle for engaging with people about the projects we are working on, a simple and succinct approach to articulate what we are doing. The second is to address people's questions and concerns, which a good story will do. The third purpose is that it makes sure that we investigate our projects thoroughly to ensure they are viable and to avoid hidden surprises when we go to implement them.

To make the process of writing the story of your project both easy and thorough there is a simple template that you can follow. The first part of the template is made up of the why, how and what of the project. Obviously this is not my own idea, it is the idea that Simon Sinek presented in his book *Start with Why*. In the book Simon points out that people are much more likely to buy your offering if you start by telling them why you do it rather than what you do.

In addition to the why, the how and the what I think there are three big questions that people have about technology. The first is, 'Is it practical?' Does it do something that matters to me and why should I care? The second is, 'Is it achievable?' Is it something that I could do (even though I don't really understand technology that much)? And the third is, 'Is it valuable?' What will it cost and what is the benefit I can expect to see in return?

If we are to smash together Simon Sinek's why, how and what and these three questions we would end up with a rather cool three by three matrix as presented in the illustration below. Investigating the project merely requires us to fill out these nine boxes. Nothing to it...

Illustration 10 - The story of our project

...well, kind of. I need to warn you that collecting all the information required to do this properly might take a little time (though this does depend on the size of the project, and how well you know the business problem and the potential solution). Patricia might have oversimplified things when she said there were three stories. Although we can broadly separate our audience into people and enablers, we will probably have multiple people, sometimes with different problems, and multiple enablers, often with different needs as well (such as IT and management). So in truth we might have to investigate and articulate many variations of our story before our work is done.

But more on that in a little bit.

enough about you, more about me

Talking about the need for personal stories seemed like a good segue into telling you my story and why I do the work I do. In 2008 I was working for Rio Tinto in my home town of Perth, Western Australia. While working there I had two different roles, the first in business improvement, which underpins a lot of my thinking about *how* we use technology better, and the second in scenario planning and strategy, which underpins my belief in *why* we need to do so.

In 2008 Rio Tinto started an experimental program to test the use of autonomous (driverless) trucks on one of its mine sites in the Pilbara. Very quickly they found out that running autonomous vehicles was not only possible, it was also incredibly profitable. It turns out that the autonomous vehicles used less fuel, reduced tyre wear, happily worked a 24 hour shift and drove safer and slower than the human driven ones. And in addition, they didn't incur the ongoing wage, accommodation and other overhead costs associated with a human driver. The program was so successful that by 2014 two of Rio Tinto's mine sites ran *only* autonomous vehicles.

Not long after the autonomous truck program started I began working in Rio Tinto Iron Ore's Scenario Planning and Strategy Team (for those not familiar with scenario planning it is an art slash science involved in taking current trends and extrapolating out multiple possible futures which in turn can be used to test strategic options). The team's job was to develop scenarios for the future with a 5-10 year time horizon, and then work with business unit managers to develop suitable strategies. One of the tools we had at our disposal was a video animation called *Mine of the Future*. In the video it showed not only a fleet of autonomous trucks, but also autonomous drill rigs, autonomous loaders, and autonomous trains. The only thing that it didn't feature a lot of was people.

When I started showing this to different business unit managers to find out what their strategy was for this autonomous future it was quite clear that no one was doing very much at all. People had so many problems to deal with in the present and were operating with such short time horizons that the challenge of automation wasn't really on their mind.

When I left Rio Tinto in 2010, after more than a decade in the workforce, I had finally found my why. My why was to help prepare people for the future of work and a future that was going to be both increasingly dynamic and incredibly technology rich. And I decided that the best way to prepare people for the future was to engage people in the technology of the present. In the words of William Gibson, "the future is already here, it's just not very evenly distributed yet".

the nine things you need to know

Let's get back to those nine boxes we need to fill in. Each box is an aspect of our project we need to investigate to ensure our project is viable. In this more detailed version of the previous illustration, presented below, there are some handy prompt questions.

The most important three questions to answer are the top three, followed by the middle and finally the ones at the bottom. It is not necessary that you fill them all out in order but if you don't get the answers to the top three questions right the rest won't matter.

We could describe this structure as 'the mullet' approach to story telling...bear with me. Anyone alive in the 1980s (or more recently if you live in Collingwood) would remember the mullet hairstyle, which involved having a short fringe and sides but keeping it long at the back. Apart from being a great reason to mock Collingwood supporters, the mullet, sometimes described as business at the front, party at the back, is a great metaphor for how we want to tell our stories.

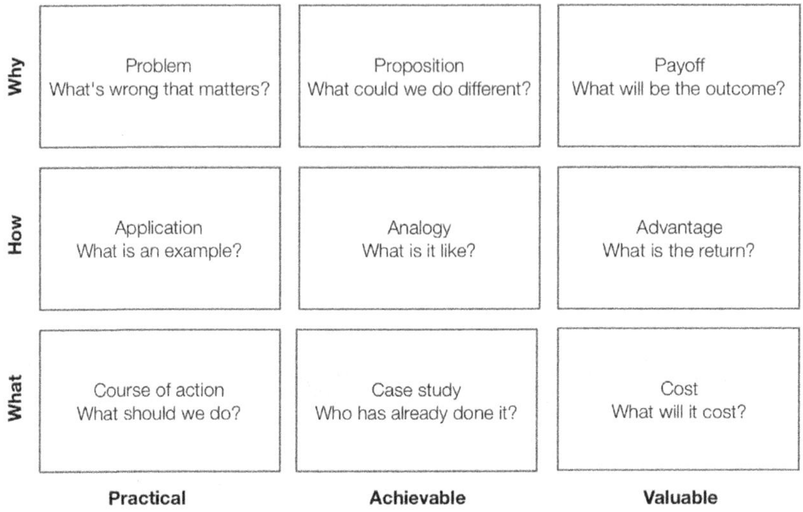

Illustration 11 - The nine things we need to know

Often when people talk about technology projects they spend lots of time talking about how the technology works, but very little talking about what it is used for. For geeky people (like me) this is the bit that we often find exciting...it's the party, but without a clear link to the business need it is pointless.

Writing our stories using the mullet approach means that we talk about the business first (the whys), the technology second (the hows) and the unruly and messy implementation bit last (the whats). The why is the ultimate motivation for the project, and getting this bit right will determine whether we get the support and resourcing we require to make our project a reality. Other parts of the project such as the how and the what are likely to change over time whereas the why will largely stay the same.

Alright then, let's get stuck in.

The three whys

The problem

The first of the whys is the problem. This is the single most important part of your story and therefore the most important answer to get right. If you are not clear about what the problem is and the pain it is causing, then pack your bags and go home. This is the key to both implementing a solution that matters to your stakeholders (rather than just another shiny object) and getting buy in from your people and enablers.

So how do you get clarity on the problem and the pain? You follow the golden rule of the digital champion: if you're not sure, then go ask. The very best problems are the ones people have told you about themselves, but even with other possibilities you have identified you can always go find the people who are potentially experiencing the problem and sense check your idea.

The best problems to solve are the painful, frustrating ones. If you can solve these then people will love you, everyone will think you're awesome and they will be falling over themselves to work with you on your next project[19]. When articulating the pain it is important to try and do this in

---

[19] Disclaimer: They may not literally fall over but it will definitely make the next project easier.

people's own words. The more personal you can make the story, the better the buy in.

The proposition

The second of the whys is the proposition. This is a statement on how our world could be fundamentally *different* than it is now (by doing the things that you're about to suggest). In some ways you could look at the proposition as the antidote to the problem identified previously (not necessarily the pain).

Unless you already know what the solution to the problem looks like, it could be difficult to articulate this future state, so although it is the second question in the table, it may not necessarily be the second question that you answer.

Perhaps the most important thing to remember about the proposition is that we are still focused on the business need and therefore we need to avoid talking about the technology. That might sound a little strange for a story about a technology project, but the focus of our story (and the focus of the project) has to be first and foremost on the business (plus there will be lots of time to talk about the technology later). In fact being able to talk about your project *without* talking about the technology is a great skill to develop.

If you're stuck with writing the proposition, a great way to get started is 'imagine if...'

The payoff

If the proposition is the antidote to the problem, the payoff is the antidote to the pain. This is a statement of how our world could be fundamentally *better* (not just different). We would all make more money, we would all work fewer hours, we would all be happier and be less stressed at work...you get the picture.

Once again, we don't really want to talk too much about the technology. We need to make it crystal clear that this is a business project, where it just so happens the solution is a digital one.

Keep the tone inspirational – this is the why and we are aiming to create desire. At this point we are not calculating the business case, we are selling the dream. Think 'save a bunch of time each week avoiding pointless meetings', not 'improve productivity by 10 hours every month'.

The three hows

While the three whys are mainly focused on the business, the three hows are mainly focused on the technology. By now your audience should be desperate to hear how you are going to solve this problem and make their lives better. Here's how we go about it.

The app

The app, or application, is your solution. It may not be an actual app (in the sense that you download it from the app store), it could be a web app, some other piece of software or even a training program that you think will help.

Whatever it is, you need to make it real.

A lot of the people you will be helping may not necessarily understand technology that well. The best way to help them understand it (and to get feedback) is to show it to them visually. It doesn't have to be the actual solution you end up using, something close will do, but the more real you can make it the more people will believe that it actually exists and you are really going to help them.

If you show how the app works on your phone/tablet/laptop and demonstrate how the proposition is achievable, you will get either buy in or feedback, or both. Buy in shows that you are on the right track, while feedback says that something in your story needs tweaking. Getting both of these things shows you have a supporter, and this is probably the best outcome of all.

The analogy

Most modern technology is a bit like black magic; it does wondrous stuff but we mostly have no idea how it works[20]. This makes it very difficult to explain it to others, and this is where an analogy comes in. A good analogy can make something we don't understand more like something we do understand. We do this by saying "it is like..." and then describing something that people are already familiar with.

It's really that simple, yet this is the one question out of the nine that people generally find hardest and often fail to answer.

---

[20] Personally I still don't understand how huge amounts of information can travel through the air using Wi-Fi and we can't see it.

A great approach to analogies is the saying, "Same, same but different". For anyone not familiar with it, you generally hear it in on the streets of Thailand or Bali, or just about anywhere else people are trying to sell you knock off goods.

"It that a real Rolex?"

"Same, same but different."

It means that in some ways what I am offering is *like* something that you are familiar with, and perhaps even desire, but there is also a difference. So, what is your solution the same as, and how is it same but different? For instance you could suggest that sending an email is the same as sending a letter, but clearly it is also different.

To create a really good analogy we want it to be the same as something people are familiar with and like and different from something people hate. For whatever reason, sending a letter still seems to have a whimsical charm attached to it, so saying that email is the same as a letter is a good start. But no one likes the fact that mail takes forever to arrive and sometimes fails to get to its destination so you might want to make it clear that email is different.

"Email is like an instantaneous letter that never fails to be delivered."

If you are still finding this bit hard, you might just need some more practice. It's like learning to ride a bike...but different.

The advantage

OK, so the business case that we avoided talking about in the why (refer to 'the payoff') appears under the advantage[21]. This is the 'what's in it for me'.

As pointed out in the previous chapter the 'what's in it for me' can be quite different for different people. My friend and mentor Matt Church looks at the different things that motivate people as currencies. Some currencies that people trade in are obvious, such as money and to a lesser extent, time. Others, such as happiness, reputation and integrity, are more obscure but often even more powerful when it comes to motivation.

When you come to calculating and articulating the advantage that your project creates you need to think about the most appropriate currency for your audience, keeping in mind that for any one project there are multiple audiences. For instance, a productivity project might serve to save each member of the executive team two hours per week and using a similar calculation as we did earlier we could estimate this to be worth $5,000+ per year. But if the executive bought into your project, did exactly what you asked and saved the two hours per week that you promised, they wouldn't see a dime of the $5,000 you promised. In this example the currency of money might be a poor motivator for the executive to get involved. For them the more appropriate currency is time. On the other hand, for the business owner who doesn't get to

---

[21] There is a good argument that this box should have been called the Payoff and the other one the Advantage, but I kind of liked that all whys started with 'P' and all the hows started with 'A'.

experience the personal benefits of the saved time, the currency is likely to be money.

The benefits of our projects are rarely one dimensional. If they save time (especially time doing work people hate) they are also likely to reduce frustration, reduce staff turnover and increase the quality of the work that is being done. Personally I think in the long run *all* of these outcomes are likely to be more valuable than the time or the money alone, but they are also almost impossible to calculate. So, just like quality needs to follow productivity when it comes to the projects we undertake, quality also needs to follow productivity when it comes to articulating the advantage.

What this means is that when articulating the advantage of our projects we need to lead with a hard currency such as time or money, and then we can follow up with the more nebulous value propositions (such as happiness and reputation) afterwards. For example:

"This project will save each of our executive team more than two hours each week, or about two full weeks each year. In addition, it will reduce stress, reduce double handling and help us do our small bit to address global warming."

Once people have bought into the hard value you can say almost anything else afterwards and they will most likely agree[22].

---

[22] I am not saying you *should* say anything afterwards, I am just saying you *can*. For our own credibility we want the value proposition to be one that we can be held accountable for and realistically achieve.

The three whats

So far, we have a clearly defined problem and possible solution. All we are missing now is some action, which comes under the three whats. This is the messy bit of our mullet.

The course of action

We never really know exactly what we need to do to implement our project. This may sound like a crazy thing to say but it is actually quite true. When people put plans in place at the beginning of a project it's not necessarily because they know exactly what needs to be done, but rather because they feel uncomfortable with the uncertainty of not knowing. All we really know is that at any one time all we have is a bunch of questions standing between us and an outcome (and at this point the outcome still might end up being that the project *doesn't* stack up and *doesn't* go ahead).

Before we move to implementation, our objective is to answer as many of these questions as we possibly can, and then have a course of action in place to answer any crucial ones that are left. As questions get answered and new ones are raised the course of action will need to change. If we were doing this exercise on paper (and I sincerely hope that you aren't) this would be the box where you would only ever work in pencil. I told you it might get a little messy.

Despite the mess, there is a certain amount of structure we can bring to devising our course of action. Firstly, we can break it down to two time frames, before pilot and after pilot. If you think the project stacks up, at some point you are going to have to run a pilot, but it makes sense to answer any questions that we possibly can first. By breaking down what we need to do before pilot and after pilot, we can focus on the right things at the right time.

Secondly, knowing that you are going to run a pilot at some point means you will also need supporters. In *The Lean Startup* (which we will talk about more under the implement section), Eric Reis talks about the importance of supporters. These are the people who want you, or more accurately, your project, to succeed. Supporters will give you honest feedback on what is working and what is not, and they will generally forgive you when things don't work. Having clarity about what you need to do next will help engage with potential supporters. The cool thing is that as a result of being engaged in the pilot supporters often become even bigger supporters because they got in on the ground floor.

The case study

Nobody wants to be the guinea pig. It will always be easier to get support for your project if someone else has done it successfully before. Being able to pull out real life examples of what other organisations have done and how they've benefited will be a great help as you tell your story. Also, through the process of investigating your project thoroughly you should have come across other people's experiences to read or hear about, so hopefully you have examples to hand.

Ideally the case studies that you have up your sleeve should be as representative of your own situation as possible. If you can, choose examples where the organisation is of similar size or in a similar industry. A great source of case studies is vendor websites. If you already have an idea of what the app or solution looks like for your project then you should be able to find case studies pretty easily.

If you had to skip over a box when telling your story it would probably be this one, but even if you do skip sharing your case studies, don't skip investigating them. You never know what you might find out and you never know when you might need them.

The cost

Wow, can you believe we are at the last box already? (I suspect you can!) Unlike some of those pesky questions we had earlier, the cost is fairly self-explanatory. What will it take to get this project off the ground? Just like the advantage, the cost is something that may need a little customisation depending on the audience. Some of the costs will be financial, which will most likely be borne by the enabler, and some of the costs will be time (such as investment in learning), which will be borne by the people using the solution. Make sure you are working in the right currency for the right people.

A couple of hints when talking about the cost.

Firstly, use the same timeframe as you did in the advantage. If you present the annual benefit per person in the advantage, present the annual cost per person as well. Secondly, if it requires the time of paid employees to implement, this doesn't normally require a dollar figure to be attached, and "a few hours over the next month" is normally an acceptable level of detail.

Finally, this is the point to think about the potential costs and advantages of hiring in some outside help. If the value proposition calculated earlier is significant enough and the complexity or time commitments to the project are something you aren't comfortable with, then this could be a good option to factor in. Of course if the value proposition is low and the project is complex, you shouldn't be hiring outside help – you shouldn't be doing the project at all (for more on this please revisit the section 'Rabbits and Rhinos').

the simplest business case in the world

When I was working in business improvement at Rio Tinto we were often required to develop a business case for the projects that we worked on. These were usually incredibly complex and took huge amounts of time, and yet I imagine they were rarely looked at in much detail. I am fairly sure that the only time they were ever closely examined was when someone wanted to find a fault in one.

I think that there are so many high value projects that digital champions could be working on that in most cases the value is self-evident, but in case it isn't (and even if it is) we want to create the simplest, most robust business case we can in the shortest amount of time.

So my formula for the simplest business case in the world is this:

- Work out how much time it will save per person each week.
- Multiply this by how many people it will help.
- Divide this by 40 (the number of hours that we pay people to work each week) which gives us an answer in terms of how many FTE, or full time employees the opportunity will save.
- We then multiple this by a rough estimate of how much we pay people. Work on $100,000 per year for an average worker and $150,000 to $200,000 a year for an executive[23], unless you have a very good reason not to (i.e. because you know a certain group of people affected by the project get paid either *substantially* higher or lower than the average Australian wage).

For instance a task tracking app that saves eight executives and 40 direct reports an hour a week by avoiding just one unnecessary meeting would calculate as:

(8 x 1) / 40 x $150,000 for the executives plus (40 x 1) /40 x $100,000 for the direct reports = $130,000 per year

Normally business cases look at the benefit over three years. If you saved $130,000 per year you could say this was a $390,000 project. Personally, I don't do that. Firstly, if it's a project worth doing the value should stack up on one year's

[23] Before you text your boss asking for a raise, this is not the *take-home* of the average employee or executive, it is the cost of employing them. It includes superannuation and overhead costs such as management costs, HR, IT and rent. They are just large, round figures that it is quite difficult for people to argue with.

returns anyway. Secondly, it is reasonable to assume that technology projects don't have as long a shelf life as infrastructure ones and projecting three solid years of returns could be a little unrealistic[24]. Finally, by sticking to one year you are reducing the chance of someone trying to pick your business case apart. If someone ever questions whether an executive is worth $130k rather than $150k you can always point out that you are only claiming one year's return and not three...when you are trying to build support for your project you don't want to spend too much time splitting hairs.

[24] This is not to say that you won't get three years of solid returns, it is just difficult to guarantee it.

two last things

There are two last things (or two first things, depending on how you look at it) to help you in sharing your project with others. You are going to need a title and an explanation.

Most people come up with incredibly boring and descriptive titles for their projects. No one really wants to be part of your project called "Rearranging the files on the shared drive". If instead your project was called "Find the things we need when we need them", or "Putting what where", you will at least earn another 30 seconds of the other person's time to tell them what the project is about.

One approach to creating a title for your project is to think how you would make it tweetable. For those of you not familiar with twitter, let me explain. The biggest issue most people have with twitter is that they can't understand how you could share anything meaningful in less than 140 characters, which is the twitter-imposed limit to how long posts can be. The truth is you can't, but what you can do in

140 characters is share a memorable headline and a hyperlink to where your audience can get more information.

The sole objective of the project title should be to get our audience to want more information. Try keeping your project titles under 140 characters and make them engaging and sharable.

For inspiration, here are a few project titles that have been shared within the Digital Champions Club recently:

- 60mins Better – a project to save construction site managers 60 minutes a week by digitising job invoices and receipts.

- The One Truth – a project to use cloud documents (such as Google Sheets) to reduce version control issues.

- Project Transparency – a project to use a collaborative task management platform to keep track of action items at the end of meetings.

If the project title is a tweet, then the project summary is a Facebook post. The project summary is for those times when you need more than a one liner and yet do not have the time or the need to go through the whole project, such as when talking to someone who isn't a potential user, enabler or supporter. In this case, it is still useful to want people to know more – who knows if they might be a potential supporter in the future.

Getting a short and succinct summary can be challenging. It reminds me of the Blaise Pascal quote, "I have made this longer than usual because I have not had time to make it shorter." Keep it to one or two sentences and address the problem, the potential solution (if you have one) and your current level of progress.

For an example please flip, click or swipe to the next section.

evaluating solutions

When it comes to evaluating different platforms, apps and solutions there are four approaches that we can take. Each of them is completely valid in their own right and deciding is about taking the right approach depending on where you are at in the project. As shown in the illustration below, these are defined as the quick, the thorough, the practical and the theoretical.

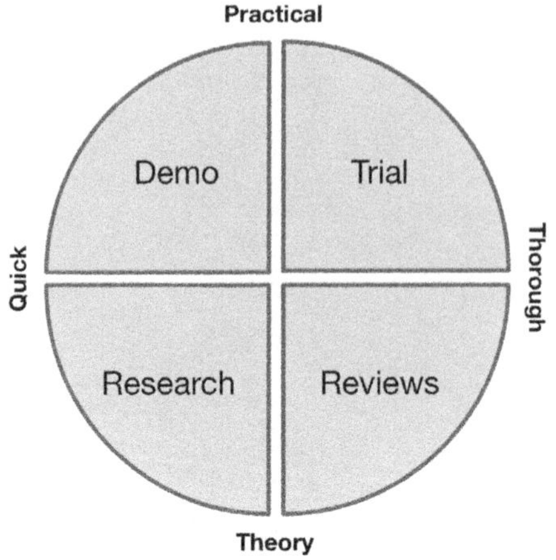

Illustration 12 - Four approaches to evaluating solutions

The starting point for any type of evaluation is research. Our objective here is to cast the net wide and capture as many potential options as possible. When looking to find a social platform for the Digital Champions Club, we considered a whole bunch of options that didn't turn out to be suitable but it was important that they were included at the beginning as it helped us shape our understanding of what we wanted. As a result we searched wikis, intranets, extranets, enterprise social and social plugins. It was only then that we found out there were a number of developers working on social plugins for the Joomla! platform, and we started finding ones that suited our needs.

During the research phase you are trying to refine your list of requirements and work out which options best meet your needs. Sometimes it isn't immediately clear, in which case the next step for me is to demo.

Most, if not all, off the shelf platforms provide some type of demo option. This might be a time limited access with sample data built in or a freemium version with limited features. When demoing the option your primary objectives are to test its usefulness (how well it meets your requirements list) and its usability (how easy and simple it is to use). You are going to need a solution that meets the requirements of IT so other considerations might be interoperability with your other platforms and systems, security, and where the data is stored and who owns it[25]. Depending on how strong your supporter base is, you might want to start showing them some of the demos to get their feedback.

As you move from investigation to implementation your evaluation needs to get a little more thorough. By now you should have narrowed down to the top five or six preferred options, and the next step is to do a deep dive on customer reviews – not just the reviews on the app store, which are normally brief and of dubious quality. Go to the support forums and online message boards and get feedback from there. If you are looking to find if specific features are available and can't get answers on the forums try emailing their support team to see if they can help (and to see how responsive they are). If you haven't already done so, this is a good point to involve IT in helping you narrow down your options.

Finally, we move onto the trial. This is when we start putting our own data and information in and seeing what comes out. By now we are probably down to a top two or three options and you may need to run a couple of different trials before

---

[25] In theory you should get some IT help and support with this, but you want to take your preferred options to IT to get their sign off rather than have them find your options.

making your final decision. Check out the chapter 'Break it down (like James Brown)' for how to limit the pain of such a trial.

As part of the trial we may seek input from supporters and some of our people to get their feedback, though we generally won't get them testing the platform for us. Ideally we want to take only one option to pilot and by this point you should know which one it is (that being said, we might have a couple of other potential solutions tucked safely away in our back pocket just in case our preferred option doesn't pan out).

extraverts and introverts

The whole premise of the digital champion is based on connecting business problems with technology solutions, but how we investigate these two things, the problems and the solutions, is fundamentally different. Investigating the problem is a very extraverted, people focused activity; there is no better way to do it than speaking to people. On the other hand, investigating the solution is a more introverted activity, involving research, testing and more attention to detail.

The theory of introversion and extraversion suggests that each person has a preference for one or the other. Specifically, one of these two approaches energises us and the other depletes us (more recently a third option, ambiversion, has been added to the mix. Apparently ambiverts are just as comfortable, or uncomfortable, in both situations). I think it is very important for digital champions to acknowledge where they sit on this spectrum as they will have a tendency to naturally prefer one type of investigation over the other. If you are aware of your bias than you can compensate a little in the other direction.

Earlier on we pointed out that digital champions could legitimately come from either the business side or the technology side of the business. It is a bit of a generalisation, but from what I've seen, those coming from the technology side tend to be a bit more introverted, whereas those from the business side tend to be a little more extraverted. One way to compensate is to have two digital champions working on projects together, one from either side of the business. This is the approach we take in the Digital Champions Club. It has the added benefit of contributing another voice to internal discussions and providing continuity should one of the digital champions leave the organisation.

here's one I prepared earlier

After you've completed your investigation, or even while you are still doing it, you need to distil everything you have learnt (and still need to learn) into a succinct and relatable story. We live in a world of too much information (or #tmi depending on which social media platform you are on) and people have neither the time nor the bandwidth to take in all this information. If you were to ever try and regurgitate verbatim all the answers to all the above questions there is a very good chance that the head of the person you are talking to will explode[26]. It is therefore incredibly important that your story is a story. Not a script, not a product brochure, but an actual story. It needs to be simple, engaging and relevant for who we are speaking to.

---

[26] I strongly advise that you don't test out this theory. Human life is precious.

Remember, our primary objective of telling our story is to *engage* our audience to a point where either a) we can get their feedback and ask them some questions or b) we can enlist their support and help in making the project happen. This means the story you tell will be subtly different depending on your audience. While 80% will probably be the same, pay particular attention to making sure that the problem, the payoff, the advantage and the cost are customised, as these are the bits that make your project relatable and valuable to your audience.

Enough talking about the story, let's look at one. Here's one I prepared earlier. This story is for the project sponsor (the person potentially paying for the project) about using the enterprise social platform Slack (slack.com). The boxes show how each element of the story relates to the nine questions we looked at earlier, but rather than read this as questions and answers, try reading out the story by just focusing on the answer boxes on the right (also, note that none of the answers are more than a couple of sentences).

Project name: Picking up the slack at ACME

Project summary

We are investigating how the use of enterprise social tools like Slack might dramatically reduce the information overload our teams are currently dealing with.

The story

The three whys

Problem (What's wrong that matters)	The amount of emails, reports and other types of information our team is dealing with just keeps going up. Our current strategy for dealing with this is to spend more time managing it which means we have less time to do the other work that matters.
Proposition (What could we do different)	What if we could find more open ways of sharing that means we didn't have to cc everyone on every email just so they can be kept informed? What if we could use a social sharing platform INSTEAD of email for internal communication?
Payoff (What will be	Our team, and especially our executive, would save time each day on managing email and have more time to spend on the

the outcome)	work that matters (such as working with their people and developing and implementing strategy).

The three hows

App (What is an example)	There is an app called Slack that allows people to communicate in open channels. Anyone can search across all of the conversations in the whole organisation to find the information or people that they need.
Analogy (What is it like)	It is kind of like Facebook for business. People can see all the conversations going on and can contribute to the ones where they have a passion or knowledge about the topic.
Advantage (What is the return)	Research by Microsoft suggests that the effective use of enterprise social platforms can lift productivity by 25%. Even if we got just a 10% increase in productivity this would be worth over $400,000 per year based on 40 people using it.

The three whats

Course of action (What we should do)	We are currently testing different platforms to see which ones might be most suitable. Then we can trial it within one team to see what they think (any ideas on who might be a suitable candidate?). Based on feedback we can then roll it out to the rest of the organisation, starting with the executive team.
Case study (Who has already done it)	I found this great case study on a similar sized organisation called OneSpaceMedia who have shared a little about how they use Slack. Let me know if you would like me to send you the link. (http://www.onespacemedia.com/news/2015/may/14/how-we-use-slack-onespacemedia/).
Cost (What will it cost)	One of the cool things is that we can test Slack for free. We don't have to pay unless we want more of the advanced features. If we do that it will be ~$80 per user per year or about $3,200 across the whole team.

Now you have a turn.

IMPLEMENT PROJECTS

think like a startup

I don't think you can talk about implementing projects in the digital space without referencing Eric Reis. His book *The Lean Startup* has become the bible for implementing projects in extreme uncertainty i.e. you don't *really* know whether there is a need for your project or even what the project will end up looking like. This is exactly what you are about to dive into. If you haven't read *The Lean Startup*, I will bear no grudges if you go and read it right now. In fact, I insist. Just bookmark this page, and I will see you in a few hours.

Back so soon?

OK, you don't have to read *The Lean Startup* right now. To save you the trouble, I am going to share some of the key takeaways pertaining to the role of a digital champion with you...I'm sure Eric won't mind.

Firstly, Eric would argue that your primary objective in a startup is not to deliver a project but to *learn*. This might sound a little strange but it does make sense. If we believe that our objective is to deliver a project then we will do our darnedest to deliver one, even if no one really wants or needs it. If on the other hand, we see that our objective is to learn, then we will run experiments to find out if our project is viable. If the project turns out to be unviable, then we still would have learnt...and not delivering a project that wasn't viable could possibly be more valuable than delivering a project that is[27].

In Eric's world there are two main types of beliefs, or hypotheses, that you want to test without experiments. The first is the value hypothesis: does this really offer value to people once they start using it? The second is the growth hypothesis: how will new customers find the product and will it scale? For a digital champion working inside an organisation, this is perhaps a little simplistic and there are two additional considerations: technological limitations and the limitations imposed by enablers.

Let me explain.

Startups don't experience the same technological limitations as an established organisation. If you are a tech startup then you have a certain ability to change or alter the technology offering available to you, or as Eric puts it, "In the modern economy almost anything that can be imagined can get built." This is fine if you have an army of computer programmers at your disposal and you are selling to a market large enough to spread the development cost. But if you are a digital champion it is likely that neither of these

---

[27] An unviable project could cost an organisation tens of thousands of dollars. NOT delivering that project could in some ways be seen as a cost saving.

things are true. You won't have the resources to hire a team of specialised programmers and you will be selling to a relatively small internal market of potential users. As a result, a digital champion is generally going to be dealing with off the shelf products, which may not always meet your exact needs.

Therefore a digital champion will most likely have a number of technology hypotheses, or beliefs about the suitability of the digital tools that are available, that will also need to be proved or disproved. For example when choosing the social platform for the Digital Champions Club we wanted the ability to 'gamify' people's contributions by using badges and rewards. Unfortunately this is not a common feature and was only available in a handful of off the shelf solutions, but because we couldn't afford to build a solution from scratch, this handful became the short list of possible options.

In addition to this, whether we like it or not, people inside organisations are generally not as open and free to make their own decisions as people on the outside. Unless a startup is operating in a regulated industry and pushing the boundaries of what is acceptable behaviour (for instance Uber in the taxi industry), it is unlikely that anyone will have either the power or desire to veto the use of the technology in a startup.

Unfortunately, finding people inside organisations with the power of veto is much, much easier. Even assuming that the growth and value hypotheses are sound, and we have both our people and the business onside, there are still many other parties who could potentially, well, spoil the party. These are subject matter experts in areas such as IT, legal, governance and other oversight roles that may have a legitimate reason to stop the project from going ahead. So in addition to the growth, value and technological hypotheses, a digital champion will also need to consider enabling hypotheses.

OK, enough about hypotheses.

If there is one word with a high probability of sending potential supporters to sleep it is hypotheses (there is also a certain social awkwardness that occurs when people try and pronounce it, especially in plural). From here on in, let's just talk about the four green lights we need to get before our project launches. And rather than talk about growth, value, technological and enabling, let's make it a little more...approachable.

Let's try it like this:

The value hypothesis = Should we do it?

The growth hypothesis = Will they use it?

The enabling hypothesis = Can we have it?

The technological hypothesis = Does it work?

So the primary objective of the digital champion during the implementation phase is to answer these four questions, and in the process of answering them, guess what? You will have launched your project.

the smudge between investigate and implement

To make the ideas in the book digestible, they have been broken down into bite sized chunks. Those chunks have been separated into chapters and the chapters grouped together as the three core activities: identify, investigate and implement. The reality of being a digital champion, and life in general, is rarely this clean. As such the line that has been drawn between what investigating and implementing is more like a smudge.

If you have got this far in your project many of the critical hypotheses have already been tested. They were tested when you showed people the app, you calculated the advantage and you went to the vendor's website to find out the features of the platform. So then, when does one end and the other begin? Well, the simplest answer is, 'when you're left with no choice'. There will be only so much that you can test and prove without launching a pilot and engaging other people.

Once again, there is a difference here between the startup world and the business world. In the startup world you can afford to engage people in a pilot and if the whole project turns south you can walk away – so in the startup world the gap between the pilot (implementation) and the investigating stage is almost non-existent. In an organisation, you need to take reasonable steps to avoid unnecessary risks and not waste other people's time and money, so a pilot should really be the first step in implementation. There is a fuzzy distinction here between testing and piloting. We are always testing but the implementation starts when we are left with no choice but to run a pilot.

get all green lights

I live in Melbourne and the traffic normally sucks, but every now and then I have an early morning flight and I'm out on the road before most of the traffic. Sometimes, just sometimes, it is possible to get green lights the whole way and not stop once between home and the airport. Most of the time, though, I will be required to stop, at least once, on the way to the airport.

A bit like this, a successful project requires that we get a green light against each one of the questions. Not just some of them, or the majority of them, all of them. Not addressing each of the questions specifically is a bit like running a red light camera...in the short term you might get to your destination sooner but it is generally not worth the long term cost and pain (and do it too many times and you might get your licence as a digital champion revoked).

One of the big mistakes that we have previously made when it comes to IT projects is that we didn't bother to get all the green lights. Instead a coercion approach was used to railroad the project through only to find out later that the project failed to meet expectations because people didn't want to use it, IT had reservations about supporting it or the value proposition didn't stack up.

So, how do we get the green lights?

Right back at the very beginning in the chapter entitled 'Connecting the dots between people, work and technology', we talked about the need for digital champions and the role they play at the centre of digital projects. Digital projects generally involve four different stakeholder groups: the executive, users, IT and vendors. What we find is that each of the four green lights we need to get align to each of these stakeholder groups. This is shown in the illustration below (which is a variation of the diagram from earlier).

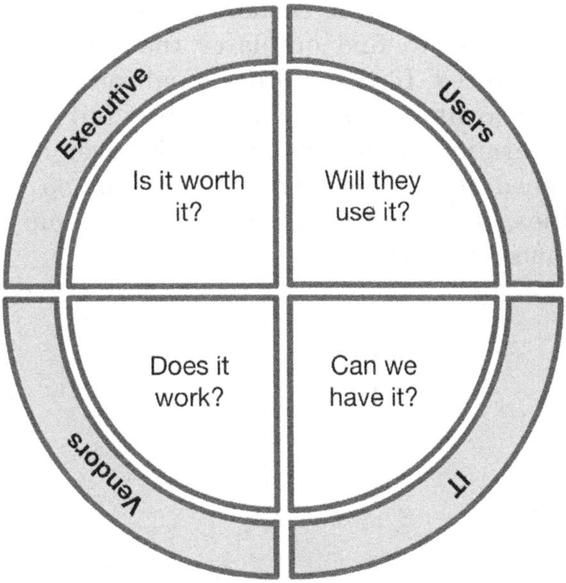

Illustration 13 - Four stakeholders and four questions

The executive or internal sponsor (who represents the business) is the person who will green light whether this is something we should do, our people will green light if this is something they want to use, IT will green light whether this is something we can do, and the vendor needs to convince us to give the app a green light for doing what we need it to do. Pretty cool, huh?

So now we have the four questions we need answered and the four stakeholders who can help answer them for us. In the next couple of sections we are going to look at how we go about doing this. But before we do that I think it is important to reiterate Eric Reis's point that the objective of the pilot is not necessarily to *succeed*, it is to *learn*. If we believe that the only objective is a successful launch, we may try and force success.

To stretch the analogy from earlier on, we may start running amber lights only to find out later that one of them had actually turned red. If on the other hand we see our primary objective as learning then we will happily fail a project that we have determined to be unviable and move on to the next one. We need to remember that there is no scarcity of opportunities, the only thing that is scarce is our time, and we want to use it well[28].

[28] Although this is true, we would still want to deliver successful projects. The value we contribute to our organisation, and subsequently the value we create for ourselves, only occurs when we implement projects that matter, the rest is just a cost.

break it down (like James Brown)

By this point you should have the following pieces of the puzzle in place:

1. A sponsor who will approve the project going ahead
2. Some enablers who you will have to satisfy as to the project's suitability
3. Some options (and probably a preference) as to what a solution might look like
4. A list of supporters who would be super excited to see your project succeed
5. A list of questions that still need answers before you can prove or disprove the value of your project.

Assuming that you have all the above then you are ready to run your pilot.

Now, this is the point where things are most likely to go pear shaped. Up until now the whole project has only existed on paper (so to speak). We have created an engaging vision, maybe demoed an example, but it has all been somewhat theoretical. As we move to the pilot stage we want to be smart about how we approach things to avoid unnecessary pain.

It is almost impossible to run a pilot, or for that matter, create any type of change without experiencing *some* pain, but just like a digital champion's sense of laziness would make them avoid long term effort, they would also want to avoid long term pain. So if a small amount of upfront pain leads to a whole lot less pain down the track then this is a good investment.

We want to approach our pilot in a way that minimises the potential for long term pain. There are three things we can do to improve our approach.

Sweat the big stuff

There are certain questions we need to answer that are bigger than others. These are the questions that if answered in the negative would kill our project in its tracks and these are the thing we want to test first. If as a result the project turns out to be unviable, we will have saved ourselves a whole bunch of effort testing other things that ultimately didn't matter.

For instance, if we are running an expense management project to speed up expense reporting, we probably need to import a bank feed into the app. Not having this feature would greatly increase the amount of data entry required and dramatically reduce the project's return, potentially making the project unviable. If this was the case then this is the feature that we would want to test first.

So, before we start piloting we want to rank our questions in terms of their risk to the project's success and get the biggest ones out the way first...don't sweat the small stuff when there is bigger stuff to be sweating over.

Break it down (like James Brown)

Sometimes bigger questions can be broken down into smaller ones and often it is advantageous to do so. By breaking down big questions we reduce the potential impact of our pilot on our people (by testing different components separately), and we get a more granular understanding of why our project will or won't work.

For example one member of the Digital Champions Club wanted to reduce the double handling and version control issues of market research data that was currently being captured in Excel spreadsheets. Their project was to test the viability of a cloud based spreadsheet solution as an alternative to Excel. One of their questions was, will our people be comfortable shifting away from Excel and using the new system? When we looked more closely we realised that there were two distinct user groups, one group collecting the data and the other involved in data analysis. We reduced the risk that conducting the pilot had on the rest of the organisation by very quickly and easily testing the functionality with the group doing the analysis without having to involve the group doing the collecting.

Pilot in parallel

For those who know a little about electronics (and trust me, I only know a little myself) there are two ways that you can wire batteries in a circuit, either in series or in parallel. Batteries in series have the positive of one connected to the negative of the next, then the positive of that battery connected to the negative of the next. This is how most of our electronics are wired and it has the advantage of increasing the total power output. The problem with it is that if one battery is faulty the whole circuit is broken and nothing will work. The alternative is to wire a circuit in parallel, whereby all the positives are wired together and all the negatives are wired together. The advantage of this is that if any of the batteries are faulty the circuit will still operate.

Running our project in parallel creates more work, as it often means getting people to participate in the pilot in addition to their day job, and working with dummy data rather than live data, but this approach will dramatically reduce the level of risk associated with your project. For instance, using the cloud based spreadsheet example from above, you would get the analysts to run their analysis using the existing Excel approach and the new cloud based solution at the same time on the same set of data. This way you can test usability, functionality and accuracy of the new system while not compromising the outcome.

As mentioned above, this experimentation serves a very important purpose. The objective is to answer our questions about the project's viability and to do so in a way that minimises the potential impact on the organisation. As you progress through the piloting process there are three potential outcomes:

- The piloting process throws up new questions and these will need to be answered before you can move forward. If this is the case, you go back to the beginning of this chapter and start again.

- The piloting process throws up a red light (a stakeholder requirement that can't be satisfactorily addressed) and you may then need to go back to the investigation phase to revisit the proposition or potentially cancel the project. If this is the case you might want to revisit the chapter 'The nine things you need to know'.

- The piloting process allows you to answer your questions satisfactorily and you are ready to proceed. If this is the case, please read on!

correcting our guesses

By now it looks like we have a fairly sound project but there is still one more thing to do. Through the piloting process we should have ended up with a green light from our people and IT on the platform or app we intend to use. Now we need our final green light: sign off from the business.

Earlier on during the investigate phase we developed a story about our project. The story was a proposition to our stakeholders and included a business case detailing the potential benefit that would accrue to both the business and supporters (the advantage) and the cost of achieving it. Before we get final sign off we need to make sure the numbers stacked up.

A warning: they won't stack up. We already know that our original value proposition is wrong because it was based on a whole bunch of assumptions that we have subsequently proved or disproved. So don't feel bad that your business case is incorrect, just figure out by how much and in what direction, then make the corrections[29].

So our objective now is not to prove our guesses correct but to correct our guesses. To do this we need to collect quantitative data (the hard numbers) from our users on the benefit that has been achieved after the pilot. This can be done either as a survey or face to face. For a small number of participants (say, fewer than 10) there are lots of good reasons to do it face to face and very few for doing it via a survey. By doing it face to face you will:

- Get a better response rate
- Get better responses
- Build relationships and rapport with your supports
- Identify and address potential concerns
- Potentially identify new digital opportunities.

---

[29] I'm making a little segue into one of my pet peeves, which is budgeting...I hope you don't mind. Budgeting is a forecast of what might happen in the future, similar to what we do during the investigation phase (but often with less thought and care). But for some reason, as soon as the budget is set it is treated as a predetermined fact.

Creating a budget and expecting those numbers to turn up is no different than thinking you can accurately predict the Lotto numbers for Saturday night. Actually, it's worse. When it comes to Lotto, after we have picked our numbers we just sit back and wait to see what happens. When we create a budget we actively manipulate outcomes to try and get it to add up...even though they were never more than a guess in the first place.

Grrrrr! Now please continue.

The advantage of doing it by survey is:

- Not having to deal with people face to face.

Finally, for small numbers there is no real time saving by using a survey. By the time you set it up, send out the survey link, send out a follow up and collate the data... you might as well have spoken to people.

What are the questions you should ask? In the following break out box there is a set of sample questions that I use when collecting data at the end of a pilot. These questions are fairly broad to enable you to gather information on benefits you might not be aware of and they also provide the opportunity to collect both 'hard data' on defined criteria (productivity, flexibility, collaboration) as well as qualitative feedback that could help you understand what our users think and whether the project has the potential to grow organically.

The focus initially is on productivity because this can be most easily translated into a dollar value and justify the cost of the project. Calculating the dollar value is clearly important when undertaking a productivity project but it is *almost* equally important when we are doing quality and positioning projects. Even if the *reason* behind the project is of higher order value (quality or positioning), it will always be easier to get the green light if you also have a financial return to *justify* the costs.

Once we have collected the data from our pilot participants it is highly likely that we will need to feed this back to our sponsor to get the green light, but we might consider feeding this back to our stakeholders as well. We all benefit from feedback and sharing the information can help in building your relationships and reputation.

some sample questions

This is a set of sample questions I use when assessing the effectiveness of programs I run and solutions I implement.

1. Has [insert app or platform name] improved your productivity?

Yes/No/Not sure

2. If you answered yes to the last question, how much time do you estimate you have saved each week?

3. In addition to productivity gains, which of the following outcomes would you also agree with (please answer Strongly Agree, Agree, Neutral, Disagree or Strongly Disagree)?

a. I feel more confident with my use of technology

b. I have increased flexibility in how and where I work

c. I feel less frustrated when using technology

d. I can collaborate more easily with colleagues

e. I better understand the opportunities and challenges of digital technology

f. I feel more effective and empowered at work

4. What other benefits or outcomes did you experience that surprised you?

5. How likely would you be to recommend [insert app or platform name] to someone else you work with (on a scale of 1–10 where 1 is highly unlikely and 10 is highly likely)?

6. Would you like any additional training or support in this area?

7. If you answered yes to the last question, what type of training and support would you like to see?

8. Any other suggestions or recommendations? We are always open to new ideas, and dare I say it, even criticism. Please feel free to share whatever else might be on your mind.

roll it out

If you have reached this point of your project then congratulations, it has been a success. Unfortunately there is little time to be resting on our laurels because we still need to make sure the rollout happens...and then there are many more projects that will need our attention.

The digital champion's role during the rollout changes significantly depending on the nature and scope of the project being undertaken. In the earlier chapter, 'Four types of projects and four different approaches', we discussed how the role of the digital champion changes as the organisation's maturity goes up. Initially they are the doer, then they become the manager, the supporter and then the mentor. This role also changes with the complexity of projects. For larger projects, especially those using outside resources to assist with deployment or training, the digital champion is more likely to take a managing role than a doing one.

This is not a book on project management so I'm not going in depth into how projects should be structured and managed. The only thing I will say about project management is that managing projects is ultimately about people. Thought leader and author Colin Ellis talks about this extensively in his book, *The Conscious Project Leader: How to create a culture of success for your projects, your people and yourself*, and I believe that this is just as true for the projects that digital champions deliver as it is for the large infrastructure projects that Colin consults on.

The role of the digital champion during this final phase of the project is therefore one that is incredibly people focused, even more so than at other times. The three key things that a digital champion needs to focus on during this time is reinforcing the purpose, maintaining engagement and filling the gaps.

Reinforcing the purpose

By this point a digital champion should be fairly invested in the project and of all the people involved it is they who best understand the underlying need. Alignment of the project with the 'why' of the project developed during the investigation phase is the ultimate test of whether the project is on track. During rollout when more people get involved it is easy for this message to get forgotten, and it is the digital champion's responsibility to keep reminding them.

Maintaining engagement

By this point the digital champion should have developed relationships with all the key stakeholders required to get the project off the ground. Even though we are in the final stages it is important for this engagement and communication to be maintained. The basis of communication might change but the need for it doesn't.

Filling the gaps

Even at this point we will find things that have been missed and it is often up to the digital champion to pick up the pieces. One of the most common requirements that is missed is training. Under the old model of IT, rollout was complete when the technology was turned on and delivered to the user, but buying and delivering the solution is just a cost. Value is only created when people are genuinely using the solution. If there are gaps between the solution being implemented and the solution being used then the digital champion needs to work out how they can help fill those gaps.

THE END OF THE BEGINNING

parting words of advice

So now we are at the end of the book I feel I should leave you with some parting words of advice. Some deeper wisdom that will carry you through the dark times.

Hmmmm...

OK, I've got something, although it's not so much parting advice as parting motivation.

A lot of this book has been focused on 'how' to be a digital champion, but never lose sight of the 'why'. Digital technology is driving significant shifts in how both our economy and our society operate. Part of the role of a digital champion is to promote a level of maturity and intentionality amongst colleagues, friends and family as to how we deal with this.

Earlier on I talked about the difference between the justification and the reason when it comes to the projects we deliver and the same thing applies to why you might want to be a digital champion. The justification might be improving the way our organisations function, driving productivity or the quality of work, or even growing our personal influence by being a catalyst of change. But underneath all this is a reason that is far more important.

Most people I meet are struggling to get good help and advice in this space and if they are left to their own devices and continue on their current path of mindless technology use the outcome might turn out pretty bleak. People are struggling to maintain a sense of balance, and are generally stuck on the challenges and not taking advantage of the benefits...of which there are many.

The best way to get people to change their approach to technology is to give them a reason to do so. Showing them what technology makes possible in a work environment will also get them to see technology differently in a personal environment. Each and every project you complete will provide a gentle nudge for people towards better technology use.

Three months ago I moved from an inner city Melbourne townhouse to live by the beach. A big part of the decision making process was in making sure my two girls grow up with a balance between the natural and the artificial, between the human and the technology. I am writing this final chapter in my studio office, in the back garden of my house. I might be in town or interstate one or two days a week depending on when and where my clients need a more 'high touch' engagement but the rest of the time I work as effectively, if not more effectively, at home than I ever did when in the city.

I jokingly talk about the move as my commitment to 'living the dream'...but it's not really a joke. The only thing that allows me to work from home, have a 20 second commute and work effectively with Sunny and Camille, my two full time staff based in the Philippines, is digital technology.

Never lose sight of the fact that digital technology is just a tool, and if we aren't using it to make our lives better, then we are using it wrong.

what's next?

At the beginning of this book I promised I would share everything I know about the role of the digital champion. As a snapshot in time of my expertise on the subject I am fairly happy with the outcome. The truth is that knowledge, and what we know to be true, is always evolving, and so unfortunately in the time between when it was printed and you got around to reading it, some of it is already out of date.

In the words of thought leader Peter Cook, "90% of everything we have ever known has turned out to be untrue.

It ultimately doesn't matter whether something is right or wrong, it only matters whether it is useful or not useful"[30].

So I hope this book is useful but I also have no doubt that at some point its usefulness will wane. Your own expertise will grow and you will start pushing at the finite boundaries of knowledge captured in the 200 odd pages of this book. This might be a point of frustration, but it is also a sign of progress.

To continue to grow and develop we all need a constant source of new knowledge and if you are a digital champion experimenting with new ideas and operating at the edge of what is already known, books aren't necessarily the best place to look. Eventually you are going to need a community of likeminded people to learn with.

You are completely welcome to create your own, and if you do, I would genuinely love to hear about it. If on the other hand you would be interested in joining mine, please check out:

digitalchampionsclub.com

or email me on simon@simonwaller.com.au

Peace out.

---

[30] I've had to paraphrase this slightly as I can't find the original quote. Knowing Pete, it was much more eloquent than that.

INDEX

Made in the USA
Monee, IL
04 November 2026

67377023R00105